Results and Problems in Cell Differentiation

Series Editors:
W. Hennig, L. Nover, U. Scheer

21

W0042545

Springer-Verlag Berlin Heidelberg GmbH

Ben A. Oostra (Ed.)

Trinucleotide Diseases and Instability

With 35 Figures

Springer

Dr. Ben A. Oostra

Department of Clinical Genetics
Erasmus University Rotterdam
Molewaterplein 50
3000 DR Rotterdam
The Netherlands

oostra@kgen.fgg.eur.nl

ISBN 978-3-662-22565-3   ISBN 978-3-540-69680-3 (eBook)
DOI 10.1007/978-3-540-69680-3

Library of Congress Cataloging-in-Publication Data

Trinucleotide diseases and instability/Ben A. Oostra, ed.
      p. cm. —(Results and problems in cell differentiation; 21)
      Includes bibliographical references and index.
      ISBN 978-3-662-22565-3
      1. Nervous system – Diseases – Genetic aspects.   2. Neurogenetics.
3. Human chromosome abnormalities.   I. Oostra, Ben A., 1946-.   II. Series.
QH607.R4 vol. 21
[RC346.4]
571.8'35 s—dc21
[616.8'0442]

© Springer-Verlag Berlin Heidelberg, 1998
Originally published by Springer-Verlag Berlin Heidelberg New York in 1998
Softcover reprint of the hardcover 1st edition 1998

Cover Design: Meta Design, Berlin
Typesetting: perform k + s textdesign GmbH, Heidelberg
SPIN 10645721   39/3137 – 5 4 3 2 1 0 – Printed on acid-free paper

Preface

Till recently, mutations in genes were described in textbooks as deletions or point mutations. These mutations can be inherited from a parent or they are de novo alterations.

The discovery in 1991 that human disease can be caused by large-scale expansion of highly unstable trinucleotide repeats has elucidated a new mutation mechanism, heritable unstable DNA. In the subsequent years more then 10 such disease genes have been identified. All dynamic mutations have been identified in neurological disorders. There are ten possible trinucleotide repeats at the DNA level, but only 3 have been identified as being involved in human diseases. The rather frequent occurence of triplet repeats in the human genome indicates that other loci subject to unstable expansions may be discovered.

The identification of repeat instability and the identification of disease genes containing trinucleotide repeats has helped to answer intriguing questions. The diseases share the unusual characteristic of inheritance with increased disease severity in successive gernerations, a phenomenon called anticipation. Trinucleotide repeat diseases are ideal subjects for direct testing because the mutation is almost exclusively of the same type and there is an extremely low occurance of new mutations in these diseases. The anticipation can now be explained by the correlation of increasing repeat length with increased disease serverity. It can be speculated that other neurological disorders showing anticipation will be caused by unstable repeats as well.

Since the identification of the unstable repeats in genes, much effort has been spent to get an answer to the following questions: What is the mechanism of the repeat instability? Why are some repeats unstable and others not? What is the function of the repeats in the (disease) genes? What are the functions of the (normal) gene products?

The different contributions cover these topics very well. The chapters describe the different groups of trinucleotide repeat genes and the mechanism of repeat instability. Although we have learned a lot in the last few years, it will be clear to the readers that there is still more exciting work to do. I hope this book will encourage them to extend their knowledge further with regard to the progress in this field.

I am grateful to all the authors in this volume for their hard work and their excellent contributions.

Rotterdam, January 1998   Ben A. Oostra

Contents

Myotonic Dystrophy
J. D. Waring and R. G. Korneluk

Instabilities of Triplet Repeats: Factors and Mechanisms
Robert D. Wells, Albino Bacolla, and Richard P. Bowater

The Fragile X Syndrome and Other Fragile Site Disorders

R. Frank Kooy[1], Ben A. Oostra[2], and Patrick J. Willems[1]

1
Introduction

Dynamic mutations are caused by amplification of a trinucleotide or other simple sequence repeat, which leads to disease and/or expression of a fragile site (Richards and Sutherland 1992). In contrast to the mendelian inheritance observed in most hereditary disorders caused by point mutations, deletions, insertions, or duplications, dynamic mutations show non-mendelian inheritance, as the amplified repeat expands or, more rarely, contracts when transmitted from one generation to the next. In the normal population, these simple sequence repeats are small, but polymorphic, having a different length. Repeats in the normal size range are stably transmitted from one generation to the next. However, a minority of individuals have larger repeats that are meiotically unstable and generally expand upon transmission from one generation to the next. If the dynamic mutation causes disease, this progressive amplification is reflected in increasing frequency and severity of the disease in successive generations. When the repeat is below a specific threshold size in an individual of the older generations of the family, the disease may not manifest itself at all. However, when the repeat exceeds a threshold size in the progeny, the disease symptoms become clear. This phenomenon of increased disease severity with successive generations, characteristic of dynamic mutations, is called anticipation.

Fragile X syndrome (Verkerk et al. 1991) and spinobulbar muscular atrophy (SBMA or Kennedy's disease; La Spada et al. 1991) were the first disorders in which a dynamic mutation was discovered. The fragile X syndrome is caused by an elongated CGG repeat in the 5' untranslated sequence of the fragile X mental retardation gene 1 (FMR1). Methylation and transcriptional silencing of the FMR1 gene follow elongation of the repeat above the threshold of approximately 200 repeats. This leads to the absence of the gene product, which causes the fragile X syndrome (Pieretti et al. 1991). Other fragile site disorders, including

[1] Department of Medical Genetics, University of Antwerp, 2610 Antwerp, Belgium
[2] Department of Clinical Genetics, Erasmus University, 3015 GE Rotterdam, The Netherlands
Correspondence to: R.F. Kooy, University of Antwerp, Department of Medical Genetics, Building T, 6th floor, Universiteitsplein 1, B-2610 Antwerp, Belgium Tel: +32-3-820.2570 Fax: +32-3-820.2566 E-mail: FKooy@UIA.AC.BE

Table 1. Dynamic diseases

Disorder	Location	Repeat	Gene	Repeat position	Repeat size in controls	Repeat size in patients	Reference
Fragile X syndrome	Xq27.3	CGG	FMR1	5' Un-translated	6–53	> 200	Verkerk et al. (1991)
FRAXE mental retardation	Xq28	CGG	FMR2	5' Un-translated	3–35	> 200	Knight et al. (1993)
Jacobsen syndrome	11q23.3	CGG	CBL2	5' Un-translated	7–25	> 100	Jones et al. (1995)
SBMA	Xq11–12	CAG	AR receptor	ORF	10–35	36– 66	La Spada et al. (1991)
Huntington's disease	4p16.3	CAG	IT-15	ORF	9–35	36–121	MacDonald et al. (1993)
SCA1	6p22–23	CAG	Ataxin-1	ORF	6–39	40– 81	Orr et al. (1993)
SCA2	12p24.1	CAG	SCA2	ORF	15–29	35– 59	Pulst et al. (1996)
Machado-Joseph's disease/SCA3	14q32.1	CAG	MJD1	ORF	7–40	61– 84	Kawaguchi et al. (1994)
SCA6	19p13	CAG	CACNL1A4	ORF	4–16	21– 27	Zhuchenko et al. (1997)
SCA7/DCA II	3p12–13	CAG	Ataxin-7	ORF	7–17	38–130	David et al (1997)
DRPLA/HRS	12pter-p12	CAG	CTG-B37	ORF	3–28	49– 75	Koide et al. (1994)
Myotonic dystrophy	19q13.3	CTG	DM kinase	3' Un-translated	3–37	> 35	Brook et al. (1992)
Friedreich's ataxia	9q13–q21.1	GAA	STM7	Intronic	8–21	> 200	Campuzano et al. (1996)

SBMA, spinal and bulbar muscular atrophy; SCA, spinocerebellar ataxia; ADCA, autosomal dominant cerebellar ataxia; DRPLA, dentatorubral pallidoluysian atrophy; HRS, Haw River syndrome; ORF, open reading frame.

FRAXE syndrome and Jacobsen syndrome, are also caused by amplified CGG repeats. These fragile site-associated disorders form the first group of dynamic diseases (Table 1). In SBMA, expansion of a CAG repeat causes elongation of the polyglutamine tract in the androgen receptor. Seven additional neurologic diseases, including Huntington's disease and several spinocerebellar ataxias, have been reported to be due to amplification of a CAG repeat in the open reading frame of the respective gene. These CAG-associated disorders form the second group of dynamic mutations. How these polyglutamine stretches cause the disease is still unknown. Myotonic dystrophy (DM), a frequent progressive syndromic muscular dystrophy, is the only disorder of a third kind of dynamic mutation caused by expansion of a CTG repeat in the 3'-untranslated part of the DM kinase gene (Brook et al. 1992). The disease mechanism is not known, but it has been suggested that the function of not only the DM kinase gene, but also that of adjacent genes is influenced by the massive CTG expansion. Consequently, each of these genes may play a role in the complex pathogenesis of DM (Harris et al. 1996). A fourth type of dynamic mutation is caused by intronic GAA triplet expansion, causing Friedreich's ataxia (Campuzano et al. 1996). In contrast to the other dynamic disorders, this neurodegenerative ataxia has an autosomal recessive mode of inheritance. Severe reduction of the Friedreich's ataxia gene transcript due to GAA repeat elongation has been suggested to be responsible for the pathology of the disease.

Over the last 7 years, 13 different dynamic disorders have been discovered, all caused by one of these four fundamentally different types of repeat expansions (for a review, see Willems 1994). This chapter will only discuss the first group of dynamic mutations caused by CGG repeat expansion, the fragile sites.

2
Fragile Sites

Fragile sites are gaps or breaks on chromosomes that arise when cells are grown under specific tissue culture conditions (Sutherland and Hecht 1985). On metaphase spreads, fragile sites are visible as discontinuities of one or both chromatin strands (Fig. 1). Fragile sites have been reported on all human chromosomes and are named according to the chromosomal band they are observed in, e.g., fra(X)(q27.3) or fra(16)(q23.2), but have also been assigned an official name in sequential order of acceptance by the HUGO nomenclature committee. For instance, the Fra(X)(q27.3) site was given the name FRAXA (fragile site, X chromosome, A site), because it was the first fragile site on the X chromosome to be recognized, and the Fra(16)(q23.2) site was given the name FRA16D (Fragile site, chromosome 16, D site), because it was the fourth recognized fragile site on chromosome 16 (Berger et al. 1985). Although most fragile sites are studied in lymphocyte cultures, expression is independent of the tissue type studied.

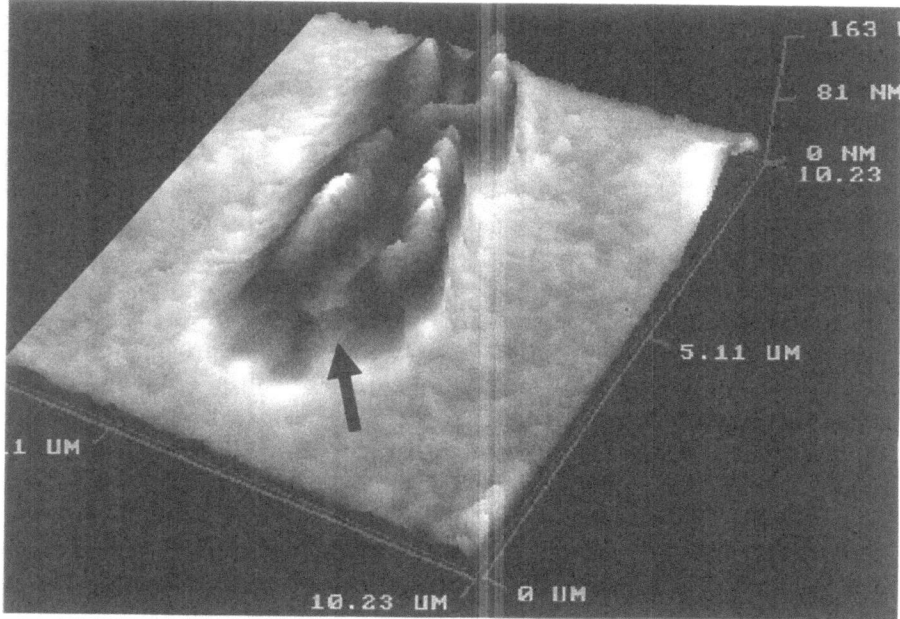

Fig. 1. A fragile X chromosome as shown by atomic force microscopy. The *arrow* indicates the fragile site

However, the percentage of cells that show fragile site expression may vary from individual to individual, a phenomenon for which there is no explanation.

Based on their relative frequency, fragile sites are divided into two types, common and rare (Berger et al. 1985; Sutherland and Hecht 1985). Common sites are possibly present in every individual, whereas rare sites have a much lower frequency. A further subdivision of fragile sites is made based on the type of chemicals necessary to induce them. The expression of most fragile sites in cells grown in standard tissue culture media is absent or very low. However, expression can be dramatically increased by cultivation in folate-deprived media or by the addition of distamycin A, bromodeoxyuridine, aphidicolin, or 5-aza-cytidine to the cell culture medium (Sutherland 1977, 1991). Table 2 shows the full classification of all fragile sites as listed in the genome data base (GDB) to date. The list is still growing, and new fragile sites are being reported on a regular basis.

2.1
Fragile Sites and Disease

Three fragile sites belonging to the rare, folate sensitive group are associated with human disorders. FRAXA is expressed in fragile X syndrome, the most frequent cause of familial mental retardation (Martin and Bell 1943; Lubs 1969).

FRAXE expression is coincident with a mild form of mental retardation in some families (Knight et al. 1993; Hamel et al. 1994). Chromosome breaks in vivo at or near the FRA11B locus cause Jacobsen syndrome, which is characterized by mental retardation and multiple malformations (Jacobsen et al. 1973; Jones et al. 1995).

Table 2. Fragile sites listed in the genome database (GDB)

Type of fragile site	Fragile sites (location)			
Rare				
Folic acid	FRA1M(1p21.3)	FRA2A (2q11.2)	FRA2B (2q13)	FRA2K (2q22.3)
	FRA5G (5q35)	FRA6A (6p23)	FRA7A (7p11.2)	FRA8A (8q22.3)
	FRA9A (9p21)	FRA9B (9q32)	FRA10A (10q23)	FRA11A (11q13.3)
	FRA11B (11q23.3)	FRA12A (12q13.1)	FRA12D (12q24.13)	FRA16A (16p13.11)
	FRA19B (19p13)	FRA20A (20p11.23)	FRA22A (22q13)	FRAXA (Xq27.3)
	FRAXE (Xq28)	FRAXF (Xq28)[a]		
Distamycin A	FRA8E (8q24.1)	FRA11I (11p15.1)	FRA16B (16q22.1)	FRA16E (16p12.1)
	FRA17A (17p12)			
BrdU	FRA10B (10q25.2)	FRA12C (12q24.2)		
Unclassified	FRA8F (8q13)			
Common				
Aphidicolin	FRA1A (1p36)	FRA1B (1p32)	FRA1C (1p31.2)	FRA1D (1p22)
	FRA1E (1p21.2)	FRA1F (1q21)	FRA1G (1q25.1)	FRA1I (1q44)
	FRA1K (1q31)	FRA1L (1p31)	FRA2C (2p24.2)	FRA2D (2p16.2)
	FRA2E (2p13)	FRA2F (2q21.3)	FRA2G (2q31)	FRA2H (2q32.1)
	FRA2I (2q33)	FRA2J (2q37.3)	FRA3A (3p24.2)	FRA3B (3p14.2)
	FRA3C (3q27)	FRA3D (3q25)	FRA4A (4p16.1)	FRA4C (4q31.1)
	FRA4D (4p15)	FRA5C (5q31.1)	FRA5D (5q15)	FRA5E (5p14)
	FRA5F (5q21)	FRA6B (6p25.1)	FRA6C (6p22.2)	FRA6E (6q26)
	FRA6F (6q21)	FRA6G (6q15)	FRA7B (7p22)	FRA7C (7p14.2)
	FRA7D (7p13)	FRA7E (7q21.2)	FRA7F (7q22)	FRA7G (7q31.2)
	FRA7H (7q32.3)	FRA7I (7q36)	FRA7J (7q11)	FRA8B (8q22.1)
	FRA8C (8q24.1)	FRA8D (8q24.3)	FRA9D (9q22.1)	FRA9E (9q32)
	FRA10D (10q22.1)	FRA10E (10q25.2)	FRA10F (10q26.1)	FRA11C (11p15.1)
	FRA11D (11p14.2)	FRA11E (11p13)	FRA11F (11q14.2)	FRA11G (11q23.3)
	FRA11H (11q13)	FRA12B (12q21.3)	FRA12E (12q24)	FRA13A (13q13.2)
	FRA13C (13q21.2)	FRA13D (13q32)	FRA14B (14q23)	FRA14C (14q24.1)
	FRA15A (15q22)	FRA16C (16q22.1)	FRA16D (16q23.2)	FRA17B (15q23.1)
	FRA18A (18q12.2)	FRA18B (18q21.3)	FRA20B (20p12.2)	FRA22B (22q12.2)
	FRAXB (Xp22.31)	FRAXC (Xq22.1)	FRAXD (Xq27.2)	
5-azacytidine	FRA1H (1q42)	FRA1J (1q12)	FRA9F (9q12)	FRA19A (19q13)
BrdU	FRA4B (4q12)	FRA5A (5p13)	FRA5B (5q15)	FRA6D (6q13)
	FRA9C (9p21)	FRA10C (10q21)	FRA13B (13q21)	
Unclassified	FRA4E (4q27)			

BrdU, bromodeoxyuridine
[a] Not yet listed in GDB.

It is not known whether additional fragile sites may cause disease. Several years ago, it was hypothesized that breakage at some sites might play a role in the chromosome rearrangements observed in many tumor cells (LeBeau and Rowley 1984; Yunis and Soreng 1984). Perhaps in line with these theories, homozygous deletions have been reported at or near the common aphidicolin-inducible FRA3B site in approximately 50 % of esophageal, stomach, and colon carcinomas. These deletions take away parts of the FHIT gene, which might prove a crucial step in the formation of digestive tract cancers (Ohta et al. 1996).

Fragile sites have been suggested to play a role in neuropsychiatric disorders, because anticipation was observed in families in which schizophrenia or bipolar affective disorder segregates (McInnis et al. 1993; Bassett and Husted 1997) and because of the observation that some patients with type I bipolar disorder or other neuropsychiatric disorders show chromosome breakage near fragile sites (Gericke 1995; Turecki et al. 1995).

Fragile sites as well as other dynamic mutations are also potential candidates for many other familial disorders in which anticipation has been observed, including spastic paraplegia, external ophthalmoplegia, or Parkinson's disease (Gispert et al. 1995; Payami et al. 1995; Melberg et al. 1996).

2.2
Molecular Base of Fragile Sites

The molecular basis of five rare, folate-sensitive fragile sites – FRAXA, FRAXE, FRAXF, FRA11B, and FRA16A (see Table 3) – consists of an elongated stretch of CGG repeats (Verkerk et al. 1991; Knight et al. 1993; Nancarrow et al. 1994; Parrish et al. 1994; Jones et al. 1995). In the normal population, this stretch is relatively limited in size (<50 repeats) and polymorphic. Chromosomes expressing the fragile site have enlarged stretches (>200 repeat units). Although many theories have been postulated, the mechanism responsible for this repeat enlargement remains unknown. Enlargement of the repeat is followed by methylation of the cytosine residues at the CpG island in which the CGG repeats are embedded. However, methylation appears not to be a prerequisite for cytogenetic expression of the fragile site, as occasional patients with fragile site expression but without CpG island methylation have been reported (McConcie-Rosell et al. 1993; Hagerman et al. 1994; Rousseau et al. 1994b; Smeets et al. 1995).

Surprisingly, identification of CGG repeats being the molecular structure of folate-sensitive fragile sites has not provided us with insight as to why, under folate-deprived conditions, fragile sites appear as nonstainable gaps in a proportion of metaphase spreads. Replication of long CGG repeats at the FRAXA locus was found to be associated with delayed replication of the cell cycle. While normal-sized repeats replicate in late S phase, expanded alleles replicate later in the cell cycle. The fragile site could therefore be the result of a disturbance

Table 3. Cloned fragile sites

Type		Site	Disorder	Location	Repeat	Stable range	Intermediate range	Cytogenetic expression	Reference
Rare	Folic acid	FRAXA	Fragile X syndrome	Xq27.3	CGG	6–53	41–230	> 200	Verkerk et al. (1991)
		FRAXE	FRAXE mental retardation	Xq28	CGG	3–35	31–200	> 200	Knight et al. (1993)
		FRAXF		Xq28	CGG	6–38	?	> 300	Parrish et al. (1994)
		FRA11B	Jacobsen syndrome	11q23.3	CGG	7–25	85–100	> 100	Jones et al. (1995)
		FRA16A		16p13.11	CGG	16–49	?	> 1000	Nancarrow et al. (1994)
	Distamycin A	FRA16B		16q22.1	ATATATTATATAT-TATATCTAATAATAT ATC/ATA	7–12	?	> 2000	Yu et al. (1997)
Common	Aphidicolin	FRA3B		3p14.2	none				Boldog et al. (1997)

of normal chromosome replication (Hansen et al. 1993, 1997; Subramanian et al. 1996). Another possible explanation for the existence of fragile sites is the existence of proteins that recognize specific trinucleotide repeats, including the CGG sequence (Richards et al. 1993; Yano-Yanagisawa et al. 1995; Deissler et al. 1997). These might accumulate at the CGG repeat when it increases in length and somehow play a role in chromosome fragility. However, protein accumulation seems unlikely to be responsible for fragile site expression on its own, as most of these proteins not only bind to CGG repeats, but also to other trinucleotide repeats such as CAG repeats, and no fragile sites are seen at chromosomal sites consisting of moderately or severely expanded CAG repeats (Jalal et al. 1993).

A more likely explanation is that fragile sites might interfere with nucleosome formation, a theory postulated many years ago (Chaudhuri 1972). Nucleosomes are basic structural elements of chromosomes consisting of 146 bp DNA coiled around a histone octamer. CGG repeats of more than 50 repeats (e.g., >150 bp) were found to be excluded from nucleosome formation using *in vitro* nucleosome reconstitution, electron microscopy and competitive assembly gel retardation analysis (Wang and Griffith 1996b; Wang et al. 1996). Methylation of the CGG repeat even increased nucleosome exclusion a further twofold (Godde et al 1996; Wang and Griffith 1996a). Failure to form nucleosomes at long stretches of methylated CGG repeats would leave the DNA relatively unprotected by histones, and this could perhaps explain the poor staining and susceptibility to chromosome breakage at fragile sites. This assumption is in line with the observation that long CTG repeats, observed for instance in DM patients, do not show fragile site expression, as CTG repeats do not exclude nucleosome formation, but actually increase the efficiency of nucleosome formation (Wang et al. 1994). This assumption is also in line with the observation that the chromatin structure is altered at expanded CCG repeats (Eberhart and Warren 1996).

Elucidation of the nucleotide sequence of the distamycin A-sensitive site FRA16B demonstrated that this rare fragile site is caused by an AT-rich minisatellite, illustrating that not only CGG trinucleotide repeats can give rise to a fragile site (Table 3). Although nothing is known about the mechanism of fragile site genesis by this type of minisatellite, it is striking that the cytogenetic expression of this site is induced by chemicals such as distamycin A and netropsin that preferentially bind AT-rich DNA (Yu et al. 1997).

Other types of rare fragile sites have not been cloned, but attempts to clone the bromodeoxyuridine-requiring site FRA10B are underway. FRA10B also appears to consist of a polymorphic sequence that is longer in carriers than in controls, but the amplified sequence is not yet known (Hewett et al. 1997).

The common fragile site FRA3B does not appear to consist of a single structural entity. Breakage at this site has been induced by aphidicolin in human-hamster cell hybrids containing chromosome 3. The breakpoints did not occur at a single position, but rather clustered in a region of 50–100 kb (Glover and Stein 1988). This region is rich in AT with an overrepresentation of MER and LINE repeats and an underrepresentation of Alu repeats, but it shows no evidence of a CGG or any other simple sequence repeat able to explain its fragility (Boldog et al. 1997).

3
Fragile X Syndrome

3.1
Phenotype

Fragile X syndrome is characterized by mental retardation together with various physical and behavioral abnormalities (for a review, see Hagerman 1996). Adult patients suffer from mild to severe mental retardation and may show macro-orchidism and have a long face, prominent ears, a high-arched palate, and flat feet. Behavioral problems include perseveration, poor eye contact, tactile defensiveness, hand flapping, anxiety, hyperactivity, hand biting, and aggressiveness. In boys, the behavioral symptoms are often of more diagnostic value than the physical ones which tend to develop gradually with age (Chudley and Hagerman 1987; Simko et al. 1989; Verma and Elango 1994; Merenstein et al. 1996). The phenotype of heterozygous females, who still have one functional copy of the fragile X gene FMR1, is less severe and even more variable than in males. Approximately 30–50 % of these females show no signs of intellectual impairment, and the remaining suffer from borderline to mild mental retardation. Physical characteristics are compatible with those found in males, but are less pronounced (Fryns 1986; Loesch and Hay 1988; Cronister et al. 1991; Hagerman et al. 1992; Mazzocco et al. 1993; de Vries et al. 1996). Irrespective of the mental impairment, female carriers of the mutation may have learning and language disabilities (Sobesky et al. 1994; Bennetto and Pennington 1996).

3.2
Prevalence

Before the identification of the expanded CGG repeat as the fragile X mutation, the estimation of the prevalence of the fragile X syndrome was based on the outcome of the cytogenetic test. Retesting of patients who had been diagnosed as affected before the discovery of the FMR1 mutation, revealed many false positives, and prevalence estimates have thus been reduced to 1 in 4000 to 1 in 6000 (Turner et al. 1996; de Vries 1997; Morton et al. 1997). Nevertheless, it is the most frequent cause of inherited mental retardation.

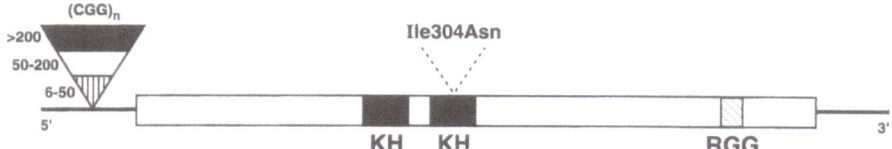

Fig. 2. FMR1 gene with KH domains and RGG boxes. In the 5' part of the gene, the CGG repeat is shown for control (6–50), premutation (50–200), and full mutation (>200)

3.3
Dynamic Mutation

The dynamic mutation leading to the fragile X syndrome consists of amplification of a CGG repeat in the 5' untranslated region (UTR) of the FMR1 gene (Fig. 2). In the normal population, the CGG repeat is polymorphic, with allele sizes ranging between six and 53 repeat units, and the repeat is inherited stably without size alteration upon transmission from generation to generation. In patients, the CGG repeat is enlarged to a full mutation consisting of a CGG allele of 200 repeats or more. Because of the repeat expansion, the promotor region of the FMR1 gene is methylated and transcriptionally silent. Hence FMR1 mRNA and FMR1 protein (FMRP) are absent in males with a full mutation (Fig. 3; Pieretti et al. 1991; Devys et al. 1993; Verheij et al. 1993; Feng et al. 1995b; de Graaff et al. 1995b; Reyniers et al. 1997). As the absence of FMRP causes the fragile X syndrome, the dynamic mutation is a loss-of-function mutation. The full mutations originate from intermediate alleles with between 41 and 200 repeats (premutations). As premutations are unmethylated, normal amounts of FMR1 mRNA and FMRP are present in premutation carriers (Feng et al. 1995b), and these individuals are therefore not affected. When transmitted from mother to progeny, the premutations may grow and expand to a full mutation. Therefore, the progeny is at risk for the fragile X syndrome (Fu et al. 1991; Devys et al. 1993). Premutations never expand to a full mutation when transmitted through

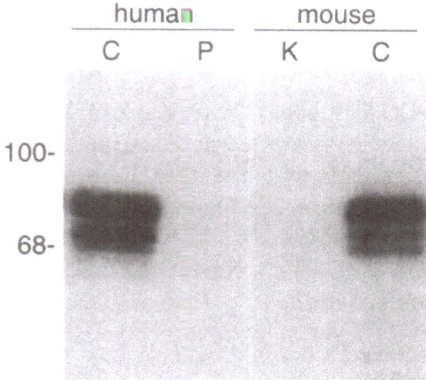

Fig. 3. FMRP expression shown by western blotting using an antibody against FMRP. Lane C, control; lane P, fragile X patient; lane K, knockout mouse

the paternal line. As a result of the dynamic inheritance of the CGG repeat, the repeat will be larger in the younger generations of a fragile X family than in previous generations, so that the younger generations in a fragile X family have a higher risk of being affected. This form of anticipation observed in fragile X families has become known as the Sherman paradox (Sherman et al. 1984, 1985).

The size of the premutation alleles overlaps partially with the normal and full mutation sized alleles (Table 2). Thus repeat size alone cannot fully distinguish the largest normal-sized alleles from the smallest premutations or the smallest full mutations from the largest premutations. In the range of 200–230 repeats, the methylation status differentiates between full mutations and premutations, as in contrast to premutations, full mutations are completely methylated. In the range of 41–53 repeats, the amount of AGG interruptions in the CGG repeat determines the stability of the alleles. Normal CGG repeats in the FMR1 gene are not perfect repeats, but are dispersed with occasional AGG triplets on average ten CGG repeats apart (Verkerk et al. 1991; Kunst and Warren 1994). AGG repeats stabilize the repeat; consequently, alleles with fewer AGG interspersions are more prone to expansion upon transmission from generation to generation, eg., as observed in the Jewish population of Tunisian origin (Falik-Zaccai et al. 1997). It has been estimated that about 34–38 uninterrupted CGG repeats are the instability threshold for FMR1 alleles (Eichler et al. 1995a).

3.4
Other Mutations

CGG repeat elongation is responsible for over 99 % of reported cases of the fragile X syndrome, but some chromosomal deletions and a few intragenic mutations inactivating the FMR1 gene have also been reported. Some of the deletions remove a whole chromosomal region from Xq27 while others take away parts of the FMR1 gene only (Fig. 4; for a review see Kooy et al. 1997). The existence of deletion mosaics and of patients with a deletion in the CGG repeat containing region only indicates that the stability of the repeat is not restricted to the CGG repeat itself, but can extend to the flanking sequences as well. These deletions in the CGG repeat region are depicted in Fig. 5, and putative hotspots for deletions around the CGG repeat have been indicated (de Graaff et al. 1995a). A frameshift mutation and a doublenucleotide substitution destroying a splice site are the only intragenic loss-of-function mutations observed (Lugenbeel et al. 1995). The phenotype of all these patients is compatible with the fragile X phenotype, once more confirming that the fragile X syndrome is the result of a loss of function of the FMR1 gene only. Additionally, two mutations in the promotor region of FMR1 have been found in two patients with mental retardation. However, the effect of these mutations on FMRP expression has only partially been characterized, and it is thus not certain whether these mutations cause disease (Mila et al. 1995).

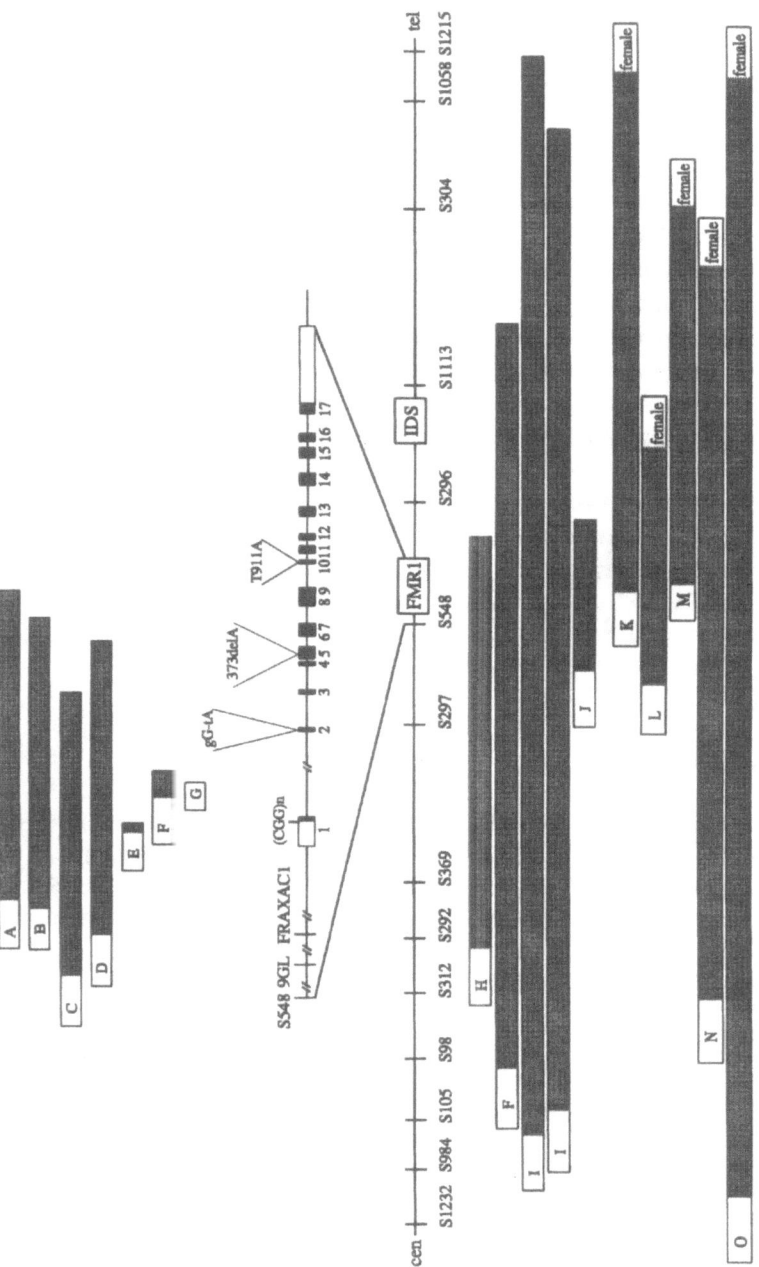

Fig. 4. Mutations in the FMR1 gene and the flanking sequences at Xq26 (not drawn to scale). Numbers 1–17 indicate the exons. Deletions are depicted as *shaded boxes* with the *letter* representing the references: A, Hirst et al. (1995); B, Gu et al. (1994); C, Wöhrle et al. (1993); D, Trottier et al. (1994); E, Meijer et al. (1994); F, Quan et al. (1995); G, Hart et al. (1995); H, Tarleton et al. (1993); I, Wolff et al. (1995); J, Gedeon et al. (1992); K, Mornet et al. (1993); L, Clarke et al. (1992); M, Dahl et al. (1995); N, Schmidt et al. (1990); O, Tharapel et al. (1993)

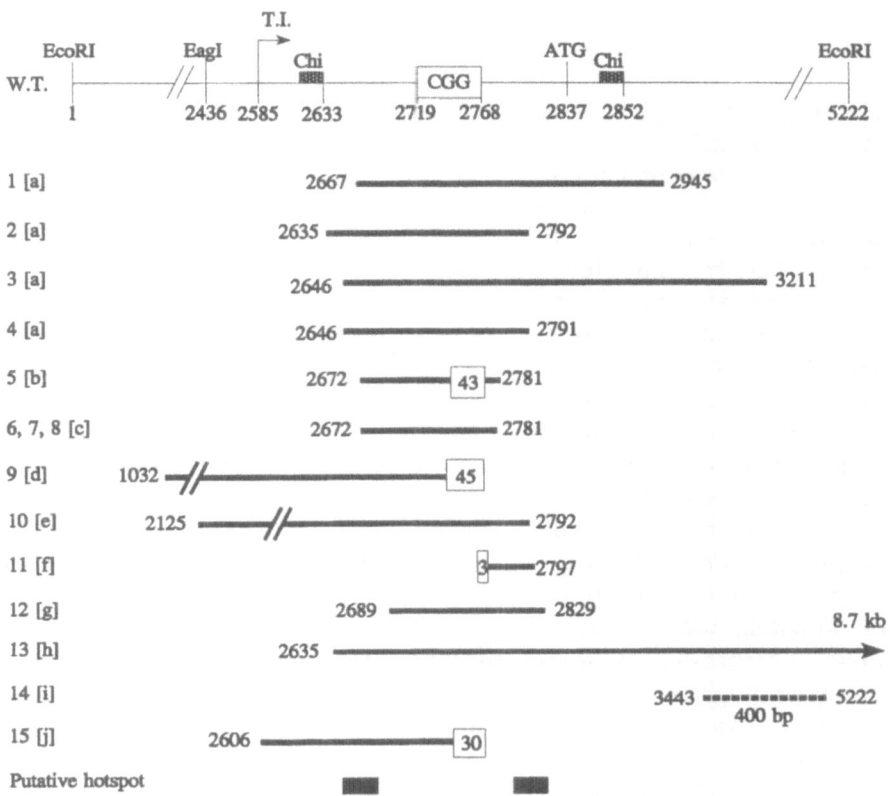

Fig. 5. Deletions in the CGG region described thus far (not drawn to scale). *Numbers* indicate the position in the pE5.1 sequence. *Boxed numbers* denote the number of CGG. WT, wildtype; T.I., translation initiation site; ATG, startcodon for translation. The deletions were obtained from the following references: a, de Graaff et al. (1995a); b, Kremer et al. (1991); c, Snow et al. (1994); d, Meijer et al. (1994); e, Hirst et al. (1995); f, de Graaff et al. (1996); g, Pulkkinen et al. (1995); h, Quan et al. (1995); i, Hart et al. (1995); j, Mila et al. (1996)

A single fragile X patient with normal amounts of FMRP has been described (De Boulle et al. 1993; Verheij et al. 1995). The patient has a point mutation in the FMR1 gene, substituting an isoleucine for an asparagine at position 304 of the coding sequence. The phenotype of this patient is peculiar, with profound mental retardation and marked macro-orchidism. The missense mutation destroys a functional RNA binding domain of the FMR1 gene, but it is not yet clear why the phenotype of this patient is more severe than that of any other fragile X patient (see Sect. 3.8).

3.5
Diagnostics

The purpose of fragile X diagnostics is to make a distinction between the three categories of CGG repeat amplification, i.e., normal-sized (6–53 repeats), intermediate sized or premutation (41–230 repeats), and full mutation alleles (> 200 repeats). The first two categories are unaffected, but in contrast to normal sized alleles, intermediate sized alleles may expand when transmitted to the offspring. Males carrying a full mutation are always affected, but about 30–50 % of female carriers of a full mutation have no detectable symptoms.

As fragile X syndrome is the most frequent cause of mental retardation, all patients referred with a mental handicap are screened as a routine procedure (Rousseau et al. 1991; Oostra et al. 1993b, Kooy et al. 1997). An initial selection is possible using PCR, allowing amplification of normal and intermediate sized alleles, but not of full mutation alleles. The presence of a normal-sized band in a male directly excludes the diagnosis of fragile X syndrome. Results other than a normal sized band do not allow a definite diagnosis to be made. If an allele in the premutation size is observed, the possibility that the patient is mosaic for the premutation and the full mutation cannot be excluded. Mosaics between a full mutation and a premutation occur frequently in males (Rousseau et al. 1994a). Absence of amplification might be caused by the presence of a full mutation, but might also be caused by poor sample quality. In females, only the presence of two separate bands in the normal size range excludes the possibility of the fragile X syndrome. If a single band is amplified, no distinction can be made between a female homozygous for the normal allele and one heterozygous for the normal allele and the full mutation.

In all cases in which PCR does not allow diagnosis, Southern blot is performed. Southern blot allows the detection of full-mutation and premutation alleles in both males and females, but is much more laborious and costly to perform (Rousseau et al. 1991; Oostra et al. 1993b; Kooy et al. 1997). Southern blotting with probe pP2 after HindIII or EcoRI digestion generates fragments of 5.2 kb in controls (Fig. 6). Premutation alleles are larger (to approximately 5.7 kb), while full-sized alleles exceed 5.7 kb. These are not normally seen as a single band, but rather as a smear of multiple bands of various intensities, indicating extensive somatic mosaicism of the CGG repeat amplification. In fact, smearing can be so intense that the full mutation is barely visible on the Southern blot and may be overlooked. Therefore, in addition to Southern blotting after HindIII or EcoRI digestion, Southern blotting after BglII digestion is performed. BglII digestion generates much larger bands of 12 kb, compressing the smear of the full mutation to a single band that is easily detected.

Southern blotting after double digestion with a methylation-sensitive enzyme such as EagI in combination with EcoRI or HindIII allows determination of the methylation status of the FMR1 promotor region. This is important in

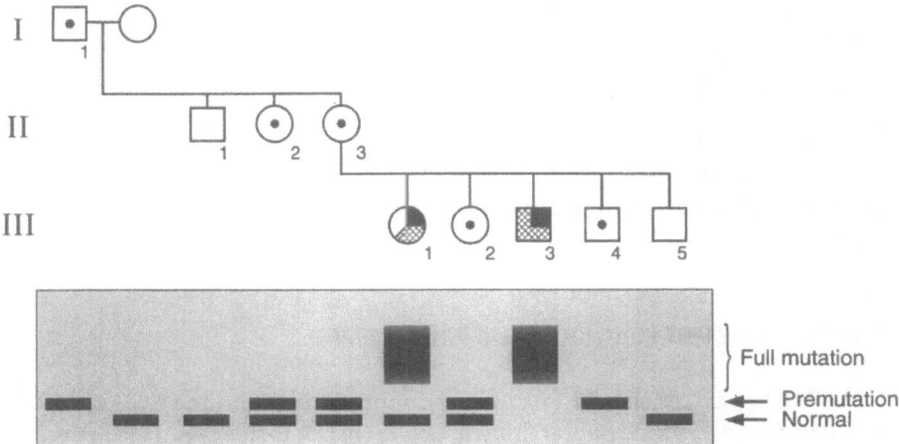

Fig. 6. DNA analysis of a fragile X family. *Square*, male; *circle*, female; *dot*, premutation *(small insert)*; *filled upper right part of symbol*, full mutation *(large insert)*; *shading in square*, intellectual disability; *shading in circle*, mild-moderate retardation in female heterozygote

discriminating the largest premutation alleles from the smallest full-mutation alleles. These sizes overlap in part, the smallest full mutation being smaller than the largest premutation, and discrimination is only possible on the base of the methylation status, as full mutations are methylated in contrast to premutations. After digestion with *Eag*I and *Hind*III, a methylated full mutation of about 200 repeats will become visible by Southern blotting as a band of approximately 5.7 kb, because the methylated promotor region is not cut by the methylation-sensitive enzyme *Eag*I. An unmethylated premutation of about the same size will be cut by *Eag*I and *Hind*III and will appear as a band of 3.2 kb. In females, because of the random X inactivation, at least two bands will be visible in controls, one inactivated by lyonization (5.2 kb) and one not inactivated (2.8 kb). A methylated full mutation in females will become visible as an extra methylated band of over 5.7 kb, and an unmethylated premutation as an extra band up to 3.2 kb.

A totally different diagnostic approach is based on direct detection of FMRP in blood cells (Willemsen et al. 1995). Smears are made from patient blood and visualized by an immunohistochemical reaction with the aid of an antibody to FMRP. Microscopic examination of the slides after the immunohistochemical reactions shows that nearly all lymphocytes are stained in control males because of the presence of the FMRP, whereas in patients no staining is detectable because of the absence of FMRP. Interestingly, as distinction by this method is based on the presence of FMRP, it will also detect other rare inactivating mutations as well as elongation and methylation of the CGG repeat. However, it is not possible to identify premutations with this technique, as these synthesize normal amounts of FMRP. One of the advantages of this technique over PCR and Southern blotting is its speed, with the result being available within hours. However, problems may be encountered when analyzing females. Due to lyonization, females have on average half of their blood cells inactivated, and hence 50 % of their cells do not generate FMRP. Extreme cases of skewed X-inactivation have been reported, and this might lead to a false-negative diagnosis in affected females, if their mutated chromosomes have been preferentially inactivated.

3.6
Fragile X Gene FMR1 and Its Protein Product FMRP

The FMR1 gene consists of 17 exons spanning 38 kb, with the CGG repeat located in the untranslated part of the first exon (Fig. 7; Eichler et al. 1993; Verkerk et al. 1993; Oostra et al. 1993a; Oostra and Willems 1995). The complete genomic sequence has been deposited in the genome sequence database (GSDB; accession no. L69074). Intronic and exonic sequences are well within the range of other proteins, except for the large 9.9-kb first exon (Eichler et al. 1993). Little is known about the FMR1 promotor. No TATA box or other promotor consensus sequence has been found in the sequences preceding the translational start codon; instead, the sequence shows the properties of a GC-rich promotor, as found, for instance, in many housekeeping genes. It has been shown, however that most of the information necessary for proper FMR1 transcription resides in 2.8 kb of sequence proximal to the first ATG, as introduction of this sequence fused to an E. Coli lacZ reporter gene in transgenic mice resulted in lacZ expression in a pattern similar to that of FMR1 in control mice (Hergersberg et al. 1995). A major transcription start site may be present about 318 bp from the initiator codon, but this has yet to be confirmed (Ashley et al. 1993a).

The FMR1 gene is subject to extensive alternative splicing in the carboxyl-terminal half of the ORF, with multiple acceptor splice sites for exons 10, 15, and 17 (Fig. 8). Moreover, exons 12 and 14 can be skipped altogether. Skipping of exon 14 alters the reading frame from exon 15 onwards, so that a protein with a different carboxyl terminus would result (Ashley et al. 1993a; Eichler et al. 1993). In the other splice variants, the reading frame is kept intact, and as a result, these all have the same carboxyl terminus. It has been calculated that up to 48 different mRNA might exist. Whether all of these are actually translated is not known, but several of these mRNA have been identified by reverse transcriptase (RT)-PCR experiments. The relative amount of each mRNA form appeared not to be dependent on the type of tissue analyzed (Verkerk et al. 1993). Translation of all splice variants of the FMR1 mRNA results in proteins of 436–631 amino acids, with a calculated molecular weight of maximally 69 kDa. On polyacrylamide gels, four to five isoforms with a molecular weight of 67–80 kDa are detected (Fig. 3). These isoforms are the products of alternative splicing, and there is no evidence for post-translational modifications (Devys et al. 1993;

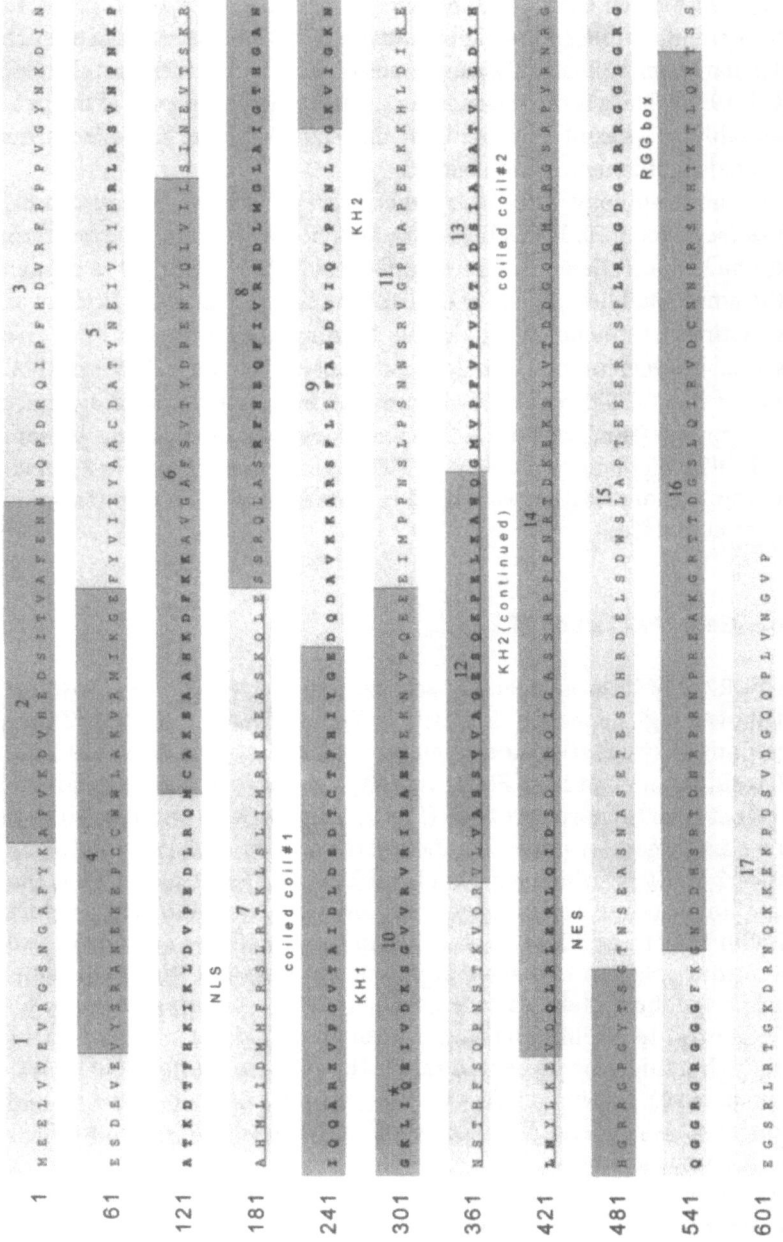

Fig. 7. FMR1 gene product, FMRP. Exons are numbered 1–17 according to Eichler et al. (1993). Structural domains are indicated in bold. Coiled coils are underlined. An asterisk indicates the isoleucine residue in the second KH domain that is mutated in the fragile X patient with the point mutation. Delineation of the structural domains is according to Musco et al. (1996) for the KH domains, Burd and Dreyfuss (1994) for the RGG box, Eberhart et al. (1996) for the NES and NLS, and Siomi et al. (1996) for the coiled coils

Siomi et al. 1993; Verheij et al. 1993; Feng et al. 1995a; Verheij et al. 1995). In addition to these high molecular weight proteins, FMRP isoforms with a much lower molecular weight (39–41 kDa) have been observed (Khandjian et al. 1995; Verheij et al. 1995). These isoforms lack the carboxyl-terminal part of the protein, presumably as a result of proteolytic cleavage. Whether these isoforms have functional significance is not clear.

FMR1 has no homology with other genes, except with its two autosomal homologues (see Sect. 3.7). It has been very well conserved during evolution, as murine, chicken and *Xenopus laevis* Fmr1 are 86–97 % identical to human FMR1 at the amino acid level (Ashley et al. 1993a; Siomi et al. 1995; Price et al. 1996). Interestingly, sequence blocks in the 3'-untranslated regions have also been conserved between human, murine, and chicken, but not *X. laevis* mRNA. Comparison of the 5'-UTR region of 44 mammalian species showed evidence of sequence conservation, with a CGG repeat, however short, being present (Deelen et al. 1994; Eichler et al. 1995b). A CGG repeat was not found in *X. laevis* or chicken. In place of a CGG repeat, chicken contains a CCT repeat (Siomi et al. 1995; Price et al. 1996).

3.7
Homologous Genes FXR1 and FXR2

FXR1 and FXR2 (FMR1 **cross**-reacting relative genes 1 and 2) are autosomal genes that show a high sequence similarity to FMR1 (Siomi et al. 1995; Zhang et al. 1995). Homology between the sequences corresponding to exons 1–13 and the first 20 residues of exon 14 of FMR1 is high, with more than 65 % of amino acids shared between the three proteins (Fig. 9). Alignment is possible without introducing a single gap, with the exception of the sequences corresponding to exons 11 and 12 of FMR1, which have not been identified in FXR1 or FXR2, and the FXR2 amino terminus, which extends for an extra ten amino acid residues compared with FMR1 and FXR1. Homology in the remainder of exon 14 and the first 14 residues of exon 15 is again apparent, although less than in the amino-terminal part of the protein. An FXR1 sequence corresponding to amino acids 445–476 in exon 14 of FMR1 has never been identified in any tested cDNA library, although it might be present in rare FXR1 isoforms as detected by RT-PCR (Coy et al. 1995). Downstream of the first residues of exon 15, FXR1 and FXR2 no longer show sequence homology to FMR1, but still share some blocks of amino acids with each other.

FXR1 mRNA is 2.2–2.4 kb in size, and the predominant protein is 70 kDa. FXR2 generates a 90-kDa protein from a 3-kb mRNA. Thus, although the coding sequences of the FXR genes correspond nearly exactly in length with that of FMR1, the mRNA are approximately half the size. This is entirely due to shorter 5'- and 3'-UTR of the FXR mRNA. Conservation of the 3'-UTR of FXR1 in evolution is remarkable, as it shows a high homology in human, mouse, and *X. laevis*, while it shows no similarity to the 3'-UTR of FMR1 or FXR2.

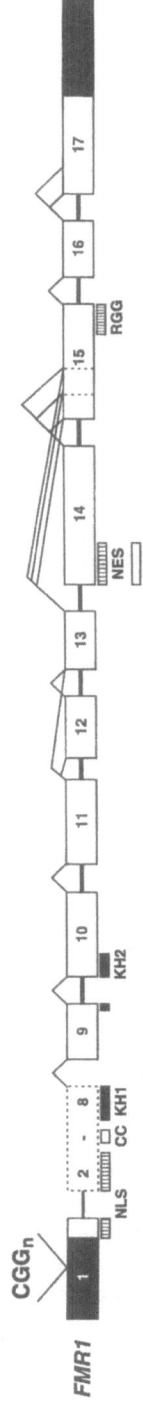

Fig. 8. Position of the domains in the FMR1 gene. NLS, Nuclear location signal; NES, Nuclear export signal; KH/RGG, KH and RGG domains important for RNA binding (see also legend to Fig. 7); cc, coiled coil region

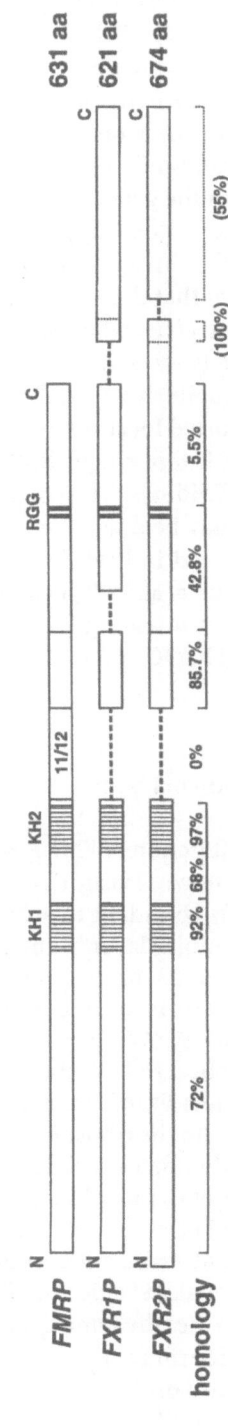

Fig. 9. Comparative position of the structural domains in FMR1 and the homologues FXR1 and FXR2. Definition of the structural domains as indicated in the legend to Fig. 7

It is not known whether the FXR genes are associated with any disorder or fragile site. Only 12 bases of the human FXR1 5'-UTR are known, and it is therefore not clear whether this 5'-UTR contains a CGG repeat. A brief $(CGG)_4$ sequence is present in the 5'-UTR of FXR2, but it has not been described whether this repeat is polymorphic in the human population or whether it can expand to a full mutation. Not unexpectedly, given the amount of fragile sites scattered throughout the genome, FXR1 and FXR2 do colocalize with fragile sites, a fact overlooked in the original description of the genes. FXR1 was originally assigned to the chromosomal region 12q12-13 (Siomi et al. 1995). However, it was later shown that the 12q12-13 gene is a processed pseudogene or an intronless form of the gene, whereas the intron-containing gene is located on chromosome 3q28 (Coy et al. 1995). FXR2 has been mapped to chromosome 17p13.1. The rare fragile sites FRA12A and FRA17A map to 12q13.1 and 17p12, at or close to the reported location of the intronless form of FXR1 and the FXR2 gene, respectively. Interestingly, both of these sites have been reported in mentally retarded individuals. The coincidence of mental retardation and fragile site expression may be due to an ascertainment bias, as both of these sites have also been observed in healthy individuals, and FRA17A even in the homozygous state (Giraud et al. 1976; Izakovic 1984; Smeets et al. 1985; Kähkönen et al. 1989). The intron-containing FXR1 gene maps close to the common aphidicolin sensitive site FRA3C at 3q27, which has not been associated with any clinical disorder.

3.8
The Function of FMRP

The exact function of FMRP, and why its absence causes mental retardation remains unknown. Tissue distribution is widespread though not ubiquitous with particularly abundant expression in the neuronal tissues of the hippocampus and the granular layer of the cerebellum (Abitbol et al. 1993; Devys et al. 1993; Hinds et al. 1993; Khandjian et al. 1995; Verheij et al. 1995). Intracellular localisation appears to be predominantly in the cytoplasm (Devys et al. 1993; Verheij et al. 1993, 1995; Willemsen et al. 1995), but occasional nuclear expression has also been observed (Verheij et al. 1993, Willemsen et al, 1996). Within the cytoplasm, most FMRP appears to be associated with the free cytoplasmic ribosomes and with the ribosomes attached to the membrane of the endoplasmic reticulum (Khandjian et al. 1996; Siomi et al. 1996; Tamanini et al. 1996; Willemsen et al. 1996), most likely with the ribosomal 60S subunit (Khandjian et al. 1996; Siomi et al. 1996). Sucrose gradient fractionation experiments suggest predominant association with the fraction of actively translating polysomes (Eberhart et al. 1996; Khandjian et al. 1996; Siomi et al. 1996). The association of FMRP to the ribosome appears to be RNA dependent, as FMRP is dissociated from the ribosomes upon RNAse treatment (Tamanini et al. 1996; Eberhart et al. 1996). Also protein-protein interactions have been reported to be important for binding of FMRP to the ribosomes (Khandjian et al. 1996; Siomi et al. 1996).

Although bound to the ribosome, FMRP is believed not to be part of its intrinsic structure, as it can be dissociated from the ribosome under relatively mild conditions (Siomi et al. 1996). It has been suggested that FMRP transports mRNA out of the nucleus (Eberhart et al. 1996). Moreover, FMRP is unlikely to bind directly to mRNAs as these bind first to the 40S subunit before binding to the 60S subunit, which binds FMRP (Khandjian et al. 1996; Willemsen et al. 1996).

FMRP has several RNA-binding domains, including two KH (heterogeneous nuclear ribonucleoprotein K homology) domains and an RGG box (Fig. 8). The single-strand RNA-binding KH domains are shared between the fragile X protein and a number of other prokaryotic and eukaryotic RNA-binding proteins as diverse as the yeast meiosis splicing factor Mer1 and vigilin (Schmidt et al. 1992; Nandabalan et al. 1993). KH domains are usually present in several copies (Burd and Dreyfuss 1994), and FMRP has two KH domains (Ashley et al. 1993b). A first clue to the elucidation of the physiological function of FMRP was presented by a patient with a very severe and atypical picture of fragile X syndrome, caused by a point mutation occurring in the second KH domain, substituting a highly conserved isoleucine residue for an asparagine residue (Fig. 2). This missense mutation reduces the ability of FMRP to bind RNA *in vitro* (De Boulle et al. 1993; Siomi et al. 1994; Verheij et al. 1995; Tamanini et al. 1996), perhaps as a result of destruction of the KH fold formed between the first and second α-helix of the KH module ßααßßα (Musco et al. 1996). A second RNA-binding domain in FMRP is the RGG box, a repeated arginine-glycine-glycine motif (Kiledjian and Dreyfuss 1992). This motif is mostly found in combination with other RNA-binding motifs, such as the KH domain (Burd and Dreyfuss 1994). In the fragile X protein, this motif is important for RNA binding, as truncation of the carboxyl-terminal part of FMRP does not alter the RNA-binding properties of FMRP, unless it includes the RGG box (Siomi et al. 1993). Apparently through its RNA-binding domains, FMRP is able to bind about 4 % of human brain mRNA in vitro, including its own mRNA (Ashley et al. 1993a; Siomi et al. 1993). In vivo, FMRP appears to associate to the ribosomes with the KH domains, as FMRP from the patient with the point mutation has reduced affinity for the ribosomal complex (Tamanini et al. 1996). Other RNA targets of FMRP and the role these interactions play in FMRP function remain to be determined.

FMRP is not only present in the cytoplasm, but also in the nucleus, where the protein is associated with the granular component of the nucleolus containing the maturing ribosomal precursor particles (Willemsen et al. 1996). As FMRP, with its molecular weight of 67–80 kDa, is too large to cross the nuclear envelope by diffusion, it must be transported actively into the nucleus. A nuclear localization signal (NLS) is located in the amino-terminal part of FMRP between residues 117 and 184 (Fig. 8; Eberhart et al. 1996). Although this region is rich in basic residues such as arginine and lysine, which is also the case in other NLS, the FMRP NLS sequence appears to be novel (Boulikas 1993; Panté and Aebi 1996).

FMRP also contains a nuclear export signal (NES), which is a small sequence motif consisting of four critically spaced, large hydrophobic amino acids within an otherwise hydrophilic sequence, first described in the retroviral protein HIV-1 Rev and in PKIα (Fischer et al. 1995; Wen et al. 1995). NES domains are involved in active nuclear export and drive proteins out of the nucleus that are too large to leave it by diffusion (e.g., larger than 40–60 kDa; for a review see Görlich and Mattaj 1996). An NES consensus sequence lies between FMRP residues 429 and 437 in the most proximal part of exon 14 (Fig. 8; Eberhart et al. 1996; Fridell et al. 1996). The proximal part of exon 14 is not present in all isoforms of FMRP as a result of alternative splicing. Thus some FMRP isoforms lack the NES of exon 14, and these were shown to have an entirely nuclear localization (Eberhart et al. 1996; Sittler et al. 1996; Willemsen et al. 1996). Whether these isoforms have a physiologic function is unknown.

Other FMRP functional domains include coiled-coil motifs predicted in exon 7 and exons 13/14 of FMRP on the basis of a computer program (Siomi et al. 1996). These motifs, involved in protein-protein interactions between two amphipathic α-helices, were initially described in keratin and myosin proteins (Cohen and Parry 1990). The predicted coiled coil motifs in exons 7 and 13/14 of FMRP are thought to be responsible for the interactions of FMRP with the FXR1 and FXR2 proteins and with the ribosome, respectively (Siomi et al. 1996).

In summary, it seems that nascent FMRP is imported into the nucleus through its NLS signal (Eberhart et al. 1996; Willemsen et al. 1996). However, it is unknown whether FMRP enters the nucleus as a monomer or bound to other proteins. After passing the nuclear membrane, FMRP is directed to the nucleolus, where it binds to the 60S subunit of the assembling ribosome (Eberhart et al. 1996; Fridell et al. 1996; Khandjian et al. 1996; Willemsen et al. 1996). After ribosome assembly, FMRP is exported to the cytoplasm, presumably as a result of the presence of the NES signal, with the nucleoporin-like protein Rab as a possible target (Fridell et al. 1996). Rab is involved in mediating the nuclear export of the NES containing HIV-1 Rev protein (Bogerd et al. 1995; Fritz et al. 1995), but its importance for FMRP nuclear export has yet to be elucidated (Fridell et al. 1996). After arriving in the cytoplasm, FMRP remains associated with the actively translating ribosomes (Eberhart et al. 1996; Khandjian et al. 1996; Siomi et al. 1996). When FMRP is dissociated from the ribosome in vitro, it is captured in a 240-kDa complex, suggesting complex formation, either with itself or with the homologous genes FXR1 and FXR2 (Siomi et al. 1996; Tamanini et al. 1996). It has not been established whether the FMRP may be released from the ribonuclein protein complex in vivo. If it is, a shuttle function of FMRP is likely (Fig. 10). Through its RNA-binding domains, it could thus transport specific mRNA into or out of the nucleus (Eberhart et al. 1996). This would be analogous to the HIV-1 Rev protein, which constantly shuttles between the nucleus and the cytoplasm and transports specific viral mRNA out of the nucleus (Meyer and Malim 1994). Like FMRP, HIV-1 Rev contains both an NES and an

NLS. Other examples of RNA-transporting proteins that shuttle between the nucleus and the cytoplasm include hnRNP A1 and U1A (Kambach and Mattaj 1992; Siomi and Dreyfuss 1995). The subcellular localization of FMRP in neurons seems to support a shuttle function of the protein. FMRP is found in neuronal perikarya and in the nuclear pores. Although neuronal FMRP is hardly found in axons, it is present throughout the dendritic tree, mainly at branch points, at the origin of spine necks, and in spine heads, all subcellular locations known to be enriched in ribosomes (Feng et al. 1997). It is precisely the dendrites that show pathologic abnormalities in the fragile X syndrome. Postmortem examination of brains of fragile X patients revealed the presence of thin, elongated dendritic spines together with mature, short, stubby spines on layer V pyramidal cells of the cerebral cortex (Hinton et al. 1991). More recent data from autopsy studies confirmed these data and moreover reported a greater spine density in patients (Comery et al. 1997b). Interestingly, similar observations were made in fragile X knockout mouse brain, showing the presence of thin and tortuous spines and increased spine density along apical dendrites (Comery et al. 1997a). Although these pathologic abnormalities are not pathognomic for the fragile X syndrome, but are also found in the brains of other retarded patients (Purpura 1974), FMRP might be involved in dendrite structure and function, and hence in synaptic transmission. This hypothesis is strengthened by the observation that the translation of FMRP at the polyribosomal fraction near synapses increases after neurotransmitter activation (Weiler et al. 1997).

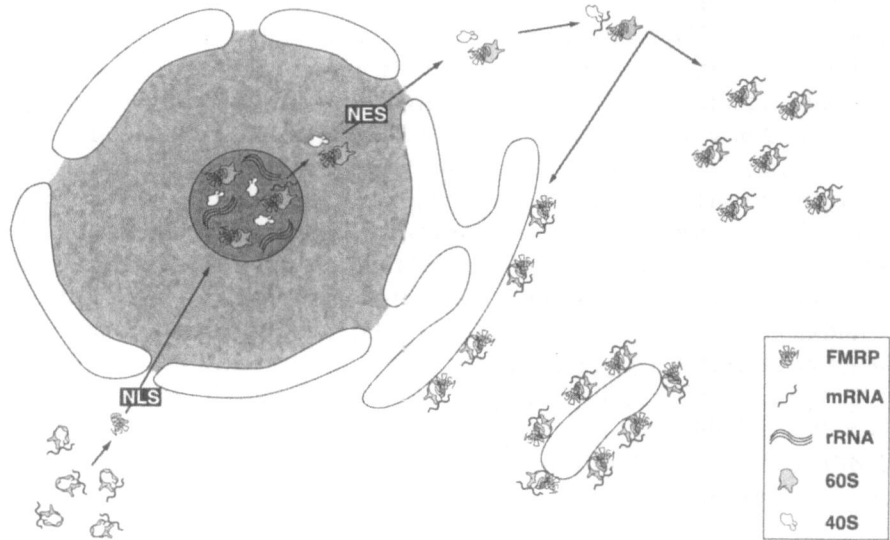

Fig. 10. Model for the intracellular routing of FMRP. NLS, Nuclear localization signal; NES, Nuclear export signal

3.9
Function of FXR1 and FXR2

The FXR1 and FXR2 proteins share many characteristics with FMRP (Fig. 9): both have a predominantly cytoplasmatic localization, similar functional domains, and a capacity to bind RNA and the 60S ribosomal subunit (Siomi et al. 1995, 1996; Coy et al. 1995; Zhang et al. 1995). Therefore, a similar function to that of FMRP can be predicted. Each of the proteins is capable of forming aggregates with each other or with itself (Zhang et al. 1995). Like FMRP, FXR1P and FXR2P are expressed in differentiated neurons of the brain only, and not in non-neuronal cells. However, in the testes, FMRP, FXR1P, and FXR2P are present in different cell types. FMRP expression is mainly detected in spermatogonia, whereas FXR1P is present in sperm cells and mature spermatogonia, and FXR2P expression is restricted to the tubuli seminiferi (Tamanini et al. 1997).

It is sometimes stated that, given the normal life span of fragile X patients and fragile X knockout mice (Bakker et al. 1994; Hagerman 1996; Kooy et al. 1996), a fundamental role for FMRP is not likely. It is indeed possible that the homologous genes FXR1 and FXR2 are capable of taking over part of the FMRP function in fragile X patients and knockout mice. The human mutations (CGG amplification or FMR1 deletion) and the knockout mutation in the transgenic mouse model are all loss-of-function mutations that lead to nearly complete absence of FMRP. Thus FMRP function could be taken over by the homologous genes in most tissues, causing a relatively mild phenotype with predominant symptoms of brain and testicular involvement both in human patients and knockout mice. The only patient with a missense mutation, an Ile-304-Asn substitution leading to a mutated form of FMRP with a nonfunctional second KH domain, has a much more severe phenotype than any other fragile X patient, with an IQ below 20 and a testicular volume of over 200 ml (De Boulle et al. 1993). This patient has normal amounts of FMRP, which is found in the same cellular compartments as in normal subjects (Verhey et al. 1995). To explain the extremely severe phenotype of this patient, one could hypothesize that the FMR1 complex with mutated FMRP containing the disrupted KH domain is less functional than a complex without FMRP at all. Thus the Ile-304-Asn mutation could cause a dominant negative loss of function, similar to the process of protein suicide. It is striking that no other patients with missense mutations in the FMR1 gene have been found, despite much research effort in several laboratories. It is possible that, in principle, these mutations disrupt the capacity of FMRP to form a complex with other proteins, and so result in severe abnormalities incompatible with life.

3.10
Repeat Expansion

The amplification of the CGG repeat in the FMR1 gene is usually determined in blood cells. Surprisingly little is known about the repeat size in other tissues. The repeat size of organs of two deceased male patients (de Graaff et al. 1995b; Reyniers et al. 1998) and a number of fetuses (Devys et al. 1992; Moutou et al. 1997) have been compared with blood and chorionic villi, and both indicate that the repeat size in all tissues is identical; however, in the male patients, an exception was found in testicular tissue, showing an additional allele of premutation size. Sperm cells and their precursors (Reyniers et al. 1993) most likely cause this additional band in testes. The presence of a premutation in the sperm of fragile X patients was anticipated because of the observation that fragile X patients may have normal daughters (Willems et al. 1992). Subsequent analysis of the sperm cells of patients confirmed the presence of an allele of premutation size only (Reyniers et al. 1993). Two models where put forward to explain where and when the expansion from a premutation to a full mutation might occur (Fig. 11). The first model suggested expansion of the repeat during maternal meiosis, with regression of the repeat during male germ cell formation. A se-

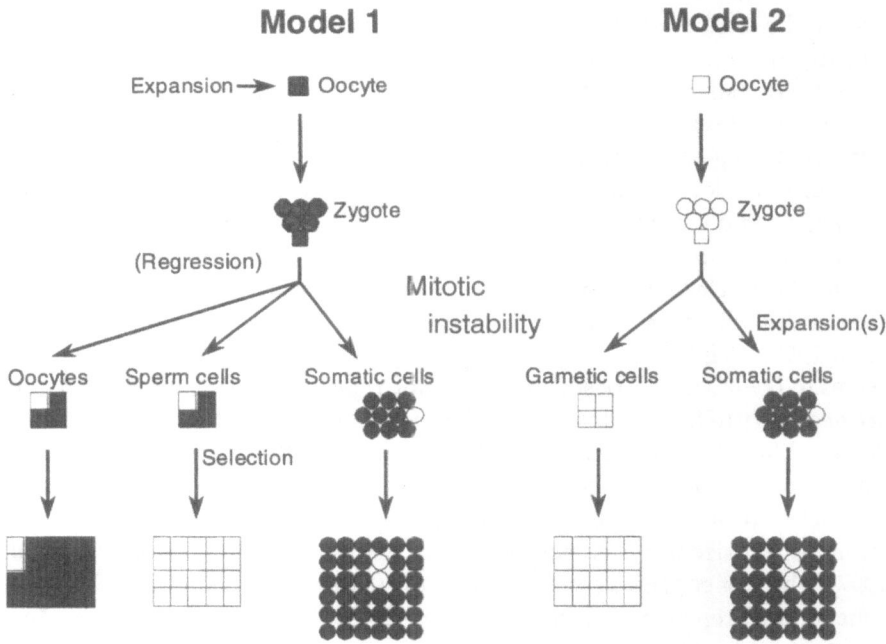

Fig. 11. Schematic depiction of the two models for the timing of repeat amplification. Somatic cells are represented by *circles*, germ cells by *squares*. Closed symbols indicate cells with a full mutation, whereas open symbols represent cells containing a premutation

lective advantage of cells with a premutation would account for the finding of only a premutation in mature sperm. A second hypothesis suggested that a full mutation sized repeat may never be present in the germ cells, but amplification occurs in early postzygotic development. Growth of the repeat would occur after the segregation of the germ cells of the embryo at day 5–6, but before the divergence of tissues and a possible twinning event at day 10–20 (Reyniers et al. 1993). This model explains both the occurrence of a premutation in the sperm as well as the similarity in repeat sizes between tissues of an individual (de Graaff et al. 1995b; Reyniers et al. 1998) or between corresponding tissues of fetal twins (Devys et al. 1992; Malmgren et al. 1992).

Discrimination between these two models is only possible if the repeat size in the oocyte is known, but direct measurements of the repeat size in single oocytes has so far been unsuccessful for technical reasons. However, examination of an ovary of a 16-week-old embryo has indicated the presence of an unmethylated full mutation in addition to a methylated one in the ovary, but not in other embryonic tissues (Malter et al. 1997). The mother of the fetus had a premutation of approximately 70 repeat units in her blood cells. This unmethylated full mutation most likely originated from the oocytes. In analogy with this observation, the testes of a 13-week-old male fetus showed evidence of a methylated full mutation, but no premutation being present, and in none of the cells in the testis could FMRP be detected. In the testes of a 17-week-old fetus, a methylated full mutation was found, but FMRP expression was shown in some cells in the testis; in contrast, in normal testes, abundant expression of FMRP was found. These findings strongly support the first hypothesis of repeat expansion during oocyte development or, alternatively, during the first few days of embryonic development before the primordial germ cells separate from the remaining embryonic cells. An occasional regression from a methylated full mutation to an unmethylated premutation in the developing male germ cells would give a selective advantage, as these cells are able to produce FMRP and as an end result only a premutation allele is found in mature sperm cells (Malter et al. 1997). While the discovery of a full mutation in early embryonic cells seems to favor a model based on premeiotic repeat expansion, additional data are necessary to fully understand CGG repeat inheritance. It is still not well understood how a carrier mother can have children with different repeat expansions. Most females show little evidence of somatic mosaicism in the blood cells, but each of the offspring that has inherited the affected chromosome has a different repeat size in his or her blood cells (Rousseau et al. 1994a; Kooy et al. 1997). This either predicts mosaicism in the female ova or, alternatively, postconceptional repeat expansion. Measurement of the repeat size in individual oocytes is obligatory in order to discriminate between these two possibilities, but it has to be mentioned that the two ovaries analyzed showed no evidence of mosaicism (Malter et al. 1997). An intriguing observation is the absence of mosaicism in the male sperm and sperm-producing cells of full-mutation males

(Reyniers et al. 1993, 1997; de Graaff et al. 1995b). This premutation must be the result of a repeat contraction from a full-sized allele back to the premutation size. In six patients with a full mutation whose allele size was determined in sperm, the CGG repeat contracted from hundreds of CGG copies to a single size between 70 and 90 repeats. This can hardly be the result of multiple sperm cells independently contracting to an identically sized repeat, but must rather be the result of regression and clonal outgrowth of a single cell. The regression could give the cell a selective advantage over the others to such an extent that it outcompetes all other cells. This advantage could be based upon the capacity of this cell to generate FMRP.

Still very little is known about the mechanism and timing of CGG repeat amplification. Several models for the repeat amplification have been proposed; the most likely model, suggested by Wells (1996) is that the expansion is the result of slippage during replication. Moreover, somatic instability also occurs. Most patients show different lengths of full mutations, often present as a smear on Southern blot analysis (Rousseau et al. 1991). This shows that repeat length changes still occur in the early embryo. In order to study the mechanism and timing of repeat instability, an animal model is required. Only in an animal model will it be possible to study gametogenesis and early embryogenesis at specific time points. To study the behavior of a premutation allele in subsequent generations, transgenic mice with a premutation allele with a repeat of $(CGG)_{81}$ in the *FMR1* promotor were generated (Bontekoe et al. 1997). Three independent lines, in total 263 transgenic animals, were tested for repeat instability. In all cases of meiosis and mitosis tested, the repeat was inherited stably. A similar result was found by Lavedan et al. (1997). In the murine *Fmr1*, CGG repeats of 9 to 12 repeat units have been identified in different mice strains. It is not known whether CGG repeat instability in mice exists, but it was assumed that underlying mechanisms causing instability would be present in humans and mice. The size of the CGG repeat necessary to cause instability in mice is not known, but might differ between humans and mice.

3.11
Animal Model

No naturally occurring animal models for the fragile X syndrome have been described, and therefore a transgenic mouse model was constructed by introduction of a neomycin cassette in exon 5 of the murine Fmr1 gene (Bakker et al. 1994). The interrupted Fmr1 gene was transfected into embryonic stem (ES) cells, and recombinant ES clones resulting from a homologous recombination event were injected into blastocysts and transferred to pseudopregnant females. Chimeric litter were crossed with wild-type females to produce females heterozygous for the knockout mutation (Bakker et al. 1994). Even though the nature of the mutation is different from most patients who have an expanded

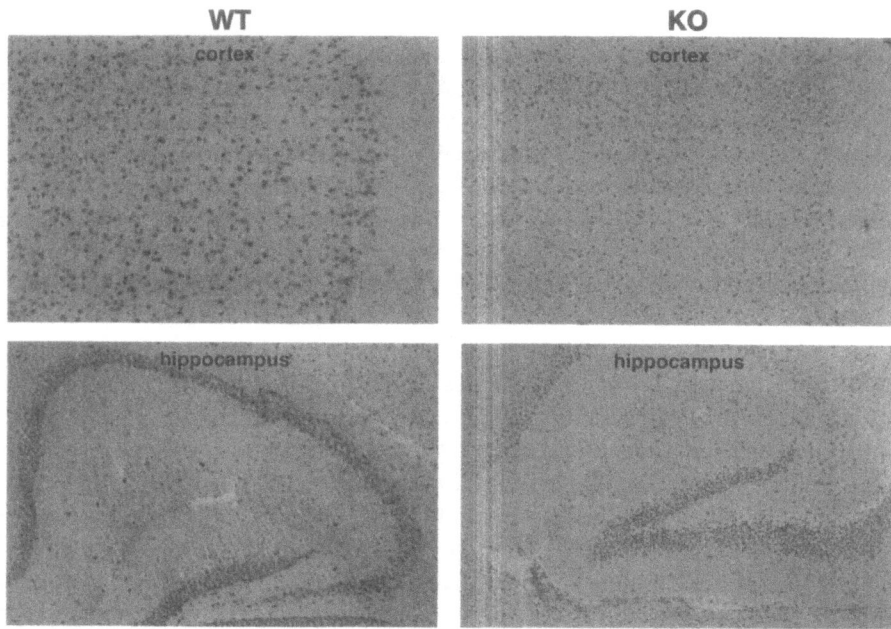

Fig. 12. Immunohistochemical staining of brain sections of wild type (WT) and knockout mice (KO), stained with antibodies against FMRP

CGG repeat in the FMR1 gene, transgenic mice, like fragile X patients, have no mature Fmr1 mRNA or protein (Figs. 2, 12; Bakker et al. 1994; Godfraind et al. 1996). The Fmr1 knockout mutation seems not to influence fertility or viability, as fragile X mice breed well and have a normal life span.

A series of tests have been performed to compare phenotypic, cognitive, behavioral, pathological, and electrophysiologic characteristics of fragile X knockout mice with their normal littermates. One of the most obvious phenotypic characteristics of fragile X males is their macro-orchidism, which sometimes manifests itself in childhood, but usually develops during puberty (Butler et al. 1992; Hagerman 1996). Macro-orchidism is found in fragile X mice from the earliest age-groups analysed and becomes more pronounced with age (Fig. 13). As in fragile X males, macro-orchidism is present in more than 90 % of adult knockout mice (Bakker et al. 1994; Kooy et al. 1996). In mice, it is entirely the result of increased Sertoli cell proliferation between embryonic day 12 and postnatal day 15 (Slegtenhorst-Eegdeman et al. 1997). Other phenotypic characteristics, such as the long face, prominent ears, high-arched palate, flat feet, hand calluses, hyperextensible finger joints, and double jointed thumb have not been found in fragile X mice (Bakker et al. 1994; Kooy et al. 1996). It should be stressed, however, that, except for measuring ear length and width, we do not have appropriate instruments to measure these parameters in mice, and mild aberrations might have gone undetected (Willems et al. 1995).

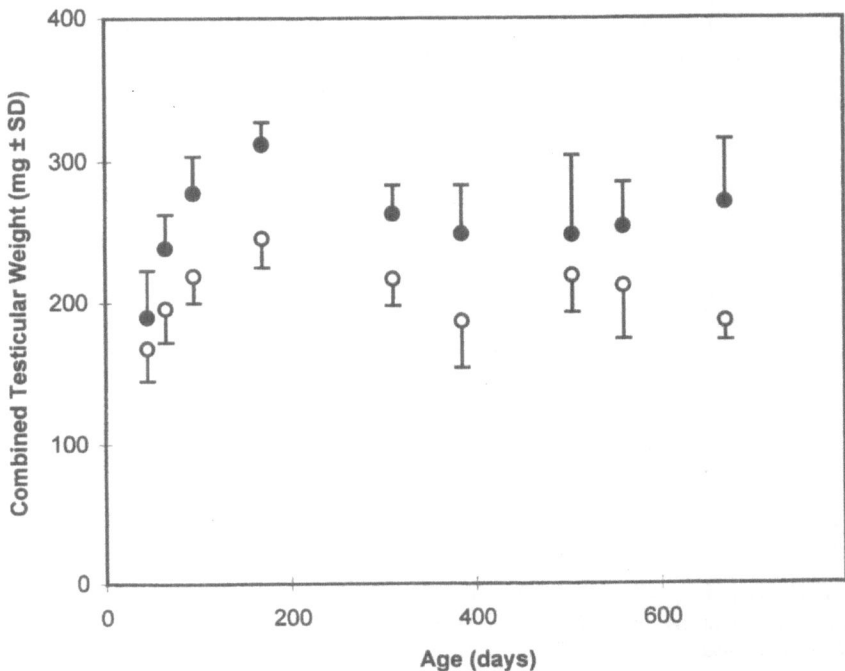

Fig. 13. Combined testicular weight of knockout (closed circles) and control mice (open circles) compared in nine age-groups

Cognitive functioning was measured in the Morris water maze test. In this test, the mice have to rely on their visual/spatial abilities to learn to find an invisible platform submerged in a circular basin filled with opaque water (Morris 1984). Fragile X patients have been reported to be deficient in visual/spatial abilities and visual short-term memory (Theobald et al. 1987; Kemper et al. 1988a, b; Cianchetti et al. 1991; Mazzocco et al. 1993; Maes et al. 1994; Bennetto and Pennington 1996). In 12 successive training sessions, the fragile X mice learned to find the hidden platform as well as the controls, although the rate of learning was marginally reduced (Fig. 14). More drastic differences were observed in the so-called reversal of the test, when the platform was placed in the opposite quadrant of the pool after the training sessions. Fragile X mice needed significantly more time to find the platform in its new position than the control littermates in all four reversal sessions (Bakker et al. 1994; Kooy et al. 1996). Using an E-maze test, the possibility was excluded that the differences in the reversal were the result of perseveration, e.g., general problems in switching from one set of learned responses to another (Kooy et al. 1996).

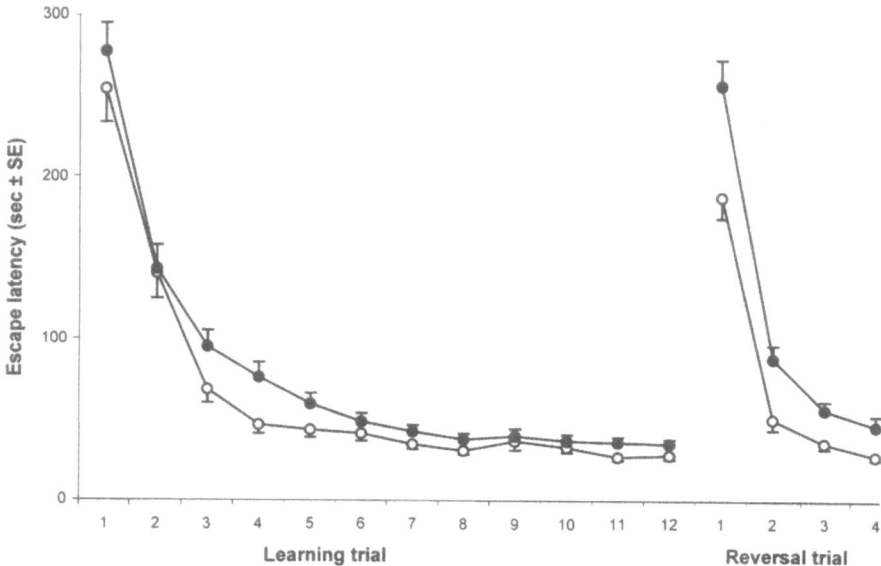

Fig. 14. Escape latency of 36 knockout mice (closed circles) and 28 control mice (open circles) in the learning and reversal trials of the Morris water maze test. Two-way ANOVA showed a significant effect of learning ($P < 0.0001$) in both phases of the trial for the two genotypes alike and an effect of genotype in both learning ($P = 0.0007$) and reversal phase of the trial ($P < 0.0001$)

Extensive statistical analysis of a subset of the Morris water maze experiments showed that the differences in escape latency were entirely the result of an increase in path length and not of a decrease in swimming speed of the fragile X mice, confirming that the differences in escape latency are most likely caused by visual/spatial disabilities and not by motor incapacity. In apparent contradiction with this, it was found that control mice did swim significantly faster in one of the control experiments, in which the normally hidden platform was made visible by a flag. As differences in swimming speed were never observed in any of the hidden platform versions of the test, the swimming speed is unlikely to have influenced the outcome of the experiments. However, the maximum swimming speed of fragile X mice when targeting a visible object seems restricted when compared to their normal littermates (D'Hooge et al. 1997).

No differences between knockout and control mice were registered in the passive avoidance test, in which mice are trained to avoid an electrical shock in the dark compartment of a step-through box. As this task only registers severe abnormalities, this result may not be unexpected (Bakker et al. 1994). In our original description of knockout mice (Bakker et al. 1994), increased locomotor activity was found in the mutant mice. Although in most subsequent experiments knockout mice had a higher motor activity than control littermates, the difference between the two genotypes was much lower than in the original experiments (Kooy et al. 1996). It now has to be determined whether the initially observed hyperactivity in knockout mice was a specific result of the Fmr1 gene knockout. Since the original description of the Fmr1 knockout mouse, transgenic mouse models for Down's syndrome and Huntington's disease have been described, and these also show increased locomotor activity, although human patients with Down's syndrome and Huntington's disease seem to display a lowered motor activity (Nasir et al. 1995; Reeves et al. 1995). It has even been suggested that hyperactivity might be a nonspecific result of the disturbance of normal neural development rather than the specific result of a single gene knockout (Sanberg et al. 1984; Moran et al. 1996).

As aberrant sizes of various brain structures, such as the cerebellar vermis, the hippocampus, and the ventricular system, of fragile X patients have been reported in magnetic resonance imaging (MRI) studies (Abrams and Reiss 1995; Reiss et al. 1995), MRI imaging of mouse brains is currently being developed to evaluate the size of these brain structures in live transgenic mice (Fig. 15; Sijbers et al. 1997). However, preliminary measurements of the vermis size do not suggest a difference between knockouts and controls.

As in fragile X patients, no pathologic abnormalities were detected in the knockout mouse using standard pathologic techniques, not even in the brain or testes, the organs most clearly affected (Bakker et al. 1994; Willems et al. 1995; Kooy et al. 1996). The only pathologic abnormality found in human fragile X brains is the presence of immature, long and tortuous spines on the dendrites in the neocortex after Golgi staining (Hinton et al. 1991; Comery et al. 1997b). Recent pathologic studies of the cortex of fragile X mice also revealed the presence of these thin and tortuous spines, whereas the spine density is elevated as compared to controls (Comery et al. 1997a). As this suggests a possible function of FMRP in maturation and pruning of neurons, and hence in synaptic transmission, we tested whether knockout mice had diminished hippocampal long-term potentiation (LTP). However, gross differences in LTP between the Fmr1 knockout mouse and controls were excluded (Godfraind et al. 1996).

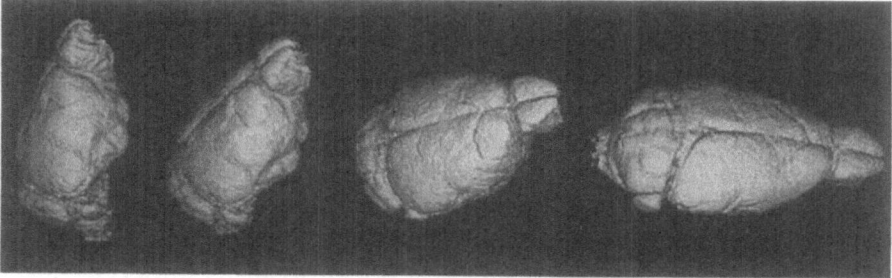

Fig. 15. 3-Dimensional MRI reconstruction of the brain of a fragile X mouse

4
FRAXE Mental Retardation

The rare folate-sensitive fragile site FRAXE located in Xq28 has only recently been recognized as being distinct from the 600 kb more proximal FRAXA in Xq27.3 (Sutherland and Baker 1992). Like the FRAXA, site the FRAXE site is caused by expansion of a CGG repeat that is hypermethylated (Knight et al. 1993). The FRAXE repeat is located in the 5'-UTR of the FMR2 gene. The entire FMR2 transcript is about 9.5 kb in length, as estimated on northern blots, and spans a genomic sequence of over 400 kb, including an exceptionally large first intron of 150 kb (Gecz et al. 1996; Gu et al. 1996). The predicted FMR2 protein contains 1276 amino acids. It has no known functional domains, but shows 28 % amino acid identity with the human MLLT2 (AF-4) gene, a gene on chromosome 4q21 found to be involved in translocations in acute lymphocyte leukemia cells (Gu et al. 1992). MLLT2 contains a putative NLS and a serine/ proline-rich domain, suggesting a possible involvement in transcriptional regulation, but its function remains unknown.

From the relatively limited number of studies reported on FRAXE families, FRAXE CGG repeat inheritance appears to parallel FRAXA repeat segregation. In the non-FRAXE-expressing population, the repeat is inherited stably and is highly polymorphic, with alleles ranging from three to 39 repeats. The smallest unstably inherited allele measured 37 repeats, and the largest unmethylated allele 230 repeats, defining the lower and upper boundary of the premutation range, respectively. Alleles larger than 270 repeats are methylated and are cytogenetically recognizable as a fragile site in all males examined and in some female carriers. Transcription of the FMR2 gene in males with an expansion more than 270 kb is undetectable (Gecz et al. 1996, 1997). Unlike the FRAXA CGG repeat, the FRAXE repeat appears not to be interrupted by occasional AGG or other non-CGG triplets (Zhong et al. 1996). During female transmission, repeat expansion is more frequent than contraction. Repeat contraction was found during male to daughter transmissions, but in contrast to the situation in the fragile X syndrome, affected males may have affected offspring (Hamel et al. 1994; Carbonell et al. 1996). Again in parallel with fragile X syndrome, the contraction may be the result of a premutation being present in the sperm of a male full mutation carrier, as observed in the single FRAXE patient whose repeat size in sperm was determined (Carbonell et al. 1996).

A relationship between FRAXE expression and mental retardation has not yet been convincingly demonstrated. In the original description of FRAXE, two families were described in which the site segregated, but no mention was made of a mental handicap cosegregating with this site (Sutherland and Baker 1992). One of these families was later shown to have expression of FRAXF, and not of FRAXE (Mulley et al. 1995). Evaluation of an extended pedigree of the second family and of a limited number of additional families suggested that FRAXE expression was being related to a mild mental handicap (Hamel et al. 1994; Knight et al. 1994; Mulley et al. 1995). It was therefore tentatively concluded that FRAXE expression might be coincident with mild to borderline mental retardation, sometimes manifested as learning difficulties or attention deficits only and without specific phenotypic abnormalities. Expressing females can be, but are not necessarily affected. In contrast to males with FRAXA expression, males with FRAXE without any detectable mental or physical handicap have been reported (Hamel et al. 1994; Murgia et al. 1996; Gecz et al. 1997).

The frequency of FRAXE mental retardation appears to be low. Large screening programs among various groups of intellectually handicapped subjects have identified very few patients (Allingham-Hawkins and Ray 1995; Biancalana et al. 1996; Holden et al. 1996). Approximately one FRAXE case was identified for each 25 FRAXA cases, allowing the conclusion that the frequency of FRAXE syndrome is approximately 4 % of that of fragile X syndrome (Brown 1996), or 1 in 100000 to 1 in 150000 in the general population. Theoretically, the number of FRAXE carriers in the normal population might be higher than estimated from screening programs of the mentally retarded, as apparently not all individuals with an elongated CGG repeat are clinically recognizable. However, screening of FRAXE allele sizes in the normal population have not revealed additional premutation- or full mutation-sized alleles, and thus such unaffected carriers cannot be common (Zhong et al. 1996; Murray et al. 1997).

In addition to the patients with an elongated CGG repeat, four mentally retarded patients with various intragenic FMR2 deletions have been described (Hamel et al. 1994; Gedeon et al. 1995; Holden et al. 1996). In only one of these patients has FMR2 gene expression been assessed and shown to be absent (Gecz et al. 1996); thus it is not known whether the other three deletions are disease-causing mutations. Expression studies of FMR2 are hampered by the fact that FMR2 mRNA is barely detectable in blood cells.

5
Jacobsen Syndrome

Jacobsen syndrome is a rare chromosome deletion syndrome involving the tip of the long arm of chromosome 11 (Jacobsen et al. 1973; Schinzel et al. 1977). Most of the 52 patients described so far have *de novo* terminal deletions involving 11q23-qter, whereas some have *de novo* interstitial deletions including band 11q24. In addition, familial cases due to balanced translocations or ring chromosomes involving 11q have been reported. The clinical presentation is variable and may include mental retardation, growth retardation, trigonocephaly, and characteristic facial dysmorphism. About 70 % of reported cases are females (Penny et al. 1995; Pivnick et al. 1996).

The CGG repeat of the rare folate-sensitive fragile site FRA11B (Voullaire et al. 1987) lies within the 5'-UTR of the proto-oncogene *CBL2* in band 11q23.3.

Parents of Jacobsen syndrome patients of three different families were shown to have moderate (85-100 copies) or even full (400-750 copies) CGG repeat expansions with fragile site expression. In the affected offspring, chromosome breakage leading to the chromosome 11q deletion had occurred near the fragile site. It was concluded that, as both FRA11B and Jacobsen syndrome are extremely rare, there must be an association between the two (Jones et al. 1995). As such, this is the first evidence of chromosome breakage at a fragile site in human patients. Breaks in FRAXA are not compatible with life. Like translocations involving FRAXA observed *in vitro* (Verkerk et al. 1991), the breaks do not occur at the CGG repeat itself, but within a chromosomal region of approximately 20 kb (Jones et al. 1995). The mechanism by which these breaks occur is still under investigation.

Whether Jacobsen syndrome is always caused by the FRA11B site is still an open question. Parents of three other Jacobsen syndrome patients have been reported to have a normal-sized CGG repeat in CBL2. Although it cannot formally be excluded that these expanded to the full mutation size in the affected offspring, this is highly unlikely from what is known about the dynamics of fragile site inheritance. In these patients, it was not possible to analyze CGG repeat length, as the deletion took away the entire *CBL2* gene, including the repeat. Chromosome breakage in these remaining patients might have occurred at other fragile sites nearby that have previously gone undetected or by a totally different mechanism. It has not been clarified, which genes cause the Jacobsen syndrome phenotype. There is no indication that the CBL2 gene is involved, as carriers of FRA11B without 11q deletions are asymptomatic, even though transcription of one copy of their *CBL2* gene is inactivated by the hypermethylation of the CpG island surrounding the CGG repeat (Jones et al. 1995).

6
Conclusion

6.1
Summary

Rare fragile sites are cytogenetically recognizable polymorphisms of human chromosomes. The molecular nature of six of these sites has recently been shown to consist of amplification of a simple repeat sequence. Like other dynamic mutations, the repeats are polymorphic in the normal population and are inherited stably. However, in carriers, the repeat is enlarged over a threshold size, resulting in expression of the fragile site and possibly in disease. Inheritance of a full mutation progeny from a normal allele has never been observed, as full mutations are derived from alleles with a repeat size between the normal and the affected range. Carriers of such a premutation are unaffected and do not show fragile site expression, but their progeny is at risk, as these premutation alleles are inherited unstably and generally expand in the next generation. It is likely that small premutation alleles only arise on very rare occasions in the normal population.

Three rare folate-sensitive fragile sites caused by amplification of a CGG repeat have been unambiguously found to be associated with disease: (1) expansion of the CGG repeat at the FRAXA locus causes fragile X syndrome, (2) expansion at the FRAXE locus is associated with a milder form of mental retardation, and (3) expansion at FRA11B is responsible for Jacobsen syndrome. Of these three disorders, the fragile X syndrome has the greatest impact in medical genetics due to its high prevalence. The fragile X gene product FMRP is an RNA-binding protein predominantly associated with the ribosomal fraction in the cytoplasm. It is possible that the protein shuttles between the nucleus and cytoplasm, thereby transporting specific mRNA into and out of the nucleus. The presence of FMRP in the neuronal dendrites, especially after neurotransmitter stimulation, suggests a possible role in synaptic transmission. Alternatively, FMRP might be essential for dendrite maturation and pruning, as immature dendrites have been observed in fragile X patients and knockout mice.

6.2
Outlook

At present it is not certain whether more fragile sites may be related to disease. Many of the fragile sites have been found in patients with a mental handicap, but this might be due to ascertainment bias, as cytogenetic analysis of mentally handicapped patients is routinely performed. A relationship between fragile sites and psychiatric illnesses has repeatedly been suggested in the past, to some extent based upon the occurrence of anticipation in affected pedigrees of various kinds of these disorders. In addition, the role of fragile sites in cancer genetics remains to be elucidated. Chromosome breaks are often observed in tumor cells, and chromosome breakage has been shown to occur near fragile sites. However, whether breakage at the fragile sites causes the chromosomal rearrangements observed in tumors remains an open question.

Acknowledgements. The contributions of the fragile X research groups in Antwerp and Rotterdam, including Edwin Reyniers, Kristel de Boulle, Isabelle Schoepen, Rudi D'Hooghe, Peter De Deyn, Jan Sijbers, Marleen Verhoye, Annemie Van Der Linden, Cathy Bakker, Carola Bontekoe, Andre Hoogeveen, Esther de Graaff, and Rob Willemsen, are greatly appreciated. We are indebted to Rudi Bernaerts for technical assistance. Financial support for fragile X syndrome research was obtained through grants from the University of Antwerp, a grant from the Belgian National Fund for Scientific Research (NFWO), from the Belgian Marguerite-Marie Delacroix foundation, from the Dutch National Fund for Scientific Research (NWO), from the American Fragile X Society, and from the European Community (BIOMED PL951663).

References

Abitbol M, Menini C, Delezoide A-L, Rhyner T, Vekemans M, Mallet J (1993) Nucleus basalis magnocellularis and hippocampus are the major sites of FMR-1 expression in the human foetal brain. Nat Genet 4:147–153

Abrams MT, Reiss AL (1995) The neurobiology of fragile X syndrome. Ment Retard Dev Dis Res Rev 1:269–275

Allingham-Hawkins DJ, Ray PN (1995) FRAXE expansion is not a common etiological factor among developmentally delayed males. Am J Hum Genet 56:72–76

Ashley CT, Sutcliffe JS, Kunst CB, Leiner HA, Eichler EE, Nelson DL, Warren ST (1993a) Human and murine FMR1: alternative splicing and translational initiation downstream of the CGG repeat. Nat Genet 4:244–251

Ashley CT, Wilkinson KD, Reines D, Warren ST (1993b) FMR-1 protein: conserved RNP family domains and selective RNA binding. Science 262:563–566

Bakker CE, Verheij C, Willemsen R, van der Helm R, Oerlemans F, Vermey M, Bygrave A, Hoogeveen AT, Oostra BA, Reyniers E, De Boulle K, D'Hooge R, Cras P, van Velzen D, Nagels G, Martin J-J, De Deyn PP, Darby JK, Willems PJ (1994) Fmr1 knockout mice: a model to study fragile X mental retardation. Cell 78:23–33

Bassett AS, Husted J (1997) Anticipation or ascertainment bias in schizophrenia? Penrose's familial mental illness sample. Am J Hum Genet 60:630–637

Bennetto L, Pennington BF (1996) The neuropsychology of fragile X syndrome. In: Hagerman RJ, Cronister A (eds) Fragile X syndrome: diagnosis, treatment, and research. Johns Hopkins University Press, Baltimore, Massachusetts, pp 210–248

Berger R, Bloomfield CD, Sutherland GR (1985) Report of the committe on chromosome rearrangements in neoplasia and on fragile sites. 8th Int Worksh on human gene mapping. Cytogenet Cell Genet 40:490–535

Biancalana V, Taine L, Bouix J-C, Finck S, Chauvin A, De Verneuil H, Knight SJL, Stoll C, Lacombe D, Mandel J-L (1996) Expansion and methylation status at FRAXE can be detected on EcoRI blots used for FRAXA diagnosis: analysis of four FRAXE families with mild mental retardation in males. Am J Hum Genet 59:847–854

Bogerd HP, Fridell RA, Madore S, Cullen BR (1995) Identification of a novel cellular cofactor for the Rev/Rex class of retroviral regulatory proteins. Cell 82:485–494

Boldog F, Gemmill RM, West J, Robinson M, Robinson L, Li E, Roche J, Todd S, Waggoner B, Lundstrom R, Jacobson J, Mullokandov MR, Klinger H, Drabkin HA (1997) Chromosome 3p14 homozygous deletions and sequence analysis of FRA3B. Hum Mol Genet 6:193–203

Bontekoe CJM, de Graaff E, Nieuwenhuizen IM, Willemsen R, Oostra BA (1997) FMR1 premutation allele is stable in mice. Eur J Hum Genet 5:293–298

Boulikas T (1993) Nuclear localisation signals (NLS). Crit Rev Eukaryot Gene Expr 3:193–227

Brook JD, McCurrach ME, Harley HG, Buckler AJ, Church D, Aburatani H, Hunter K, Stanton VP, Thirion J-P, Hudson T, Sohn R, Zemelman B, Snell RG, Rundle SA, Crow S, Davies J, Shelbourne P, Buxton J, Jones C, Juvonen V, Johnson K, Harper PS, Shaw DJ, Housman DE (1992) Molecular basis of myotonic dystrophy: expansion of a trinucleotide (CTG) repeat at the 3' end of a transcript encoding a protein kinase family member. Cell 68:799–808

Brown WT (1996) The FRAXE syndrome: is it time for routine screening? Am J Hum Genet 58:903–905

Burd CG, Dreyfuss G (1994) Conserved functions and diversity of functions of RNA binding proteins. Science 615–621

Butler MG, Brunschwig LK, Miller LK, Hagerman RJ (1992) Standards for selected anthropometric measurements in males with the fragile X syndrome. Pediatrics 89:1059–1062

Campuzano V, Montermini L, Moltò MD, Pianese L, Cossée M, Cavalcanti F, Monros E, Rodius F, Duclos F, Monticelli A, Zara F, Cañizares J, Koutnikova H, Bidichandani SI, Gellera C, Brice A, Trouillas P, De Michele G, Filla A, De Frutos R, Palau F, Patel PI, Di Donato S, Mandel J-L, Cocozza S, Koenig M, Pandolfo M (1996) Friedreich's ataxia: autosomal recessive disease caused by an intronic GAA triplet repeat expansion. Science 271:1423–1427

Carbonell P, López I, Gabarrón J, Bernabé MJ, Lucas JM, Guitart M, Gabau E, Glover G (1996) FRAXE mutation analysis in three Spanish families. Am J Med Genet 64:434–440

Chaudhuri JP (1972) On the origin and nature of achromatic lesions. Chromosomes Today 3:147–151

Chudley AE, Hagerman RJ (1987) Fragile X syndrome. J Pediatr 110:821–831

Cianchetti C, Sannio-Fancello G, Fratta A-L, Manconi F, Orano A, Pischedda M-P, Pruna D, Spinicci G, Archidiacono N, Filippi G (1991) Neuropsychological, psychiatric, and physical manifestations in 149 members from 18 fragile X families. Am J Med Genet 40:234–243

Clarke JT, Wilson PJ, Morris CP, Hopwood JJ, Richards RI, Sutherland GR, Ray PN (1992) Characterization of a deletion at Xq27-q28 associated with unbalanced inactivation of the nonmutant X chromosome. Am J Hum Genet 51:316–322

Cohen C, Parry DAD (1990) α-Helical coiled coils and bundles: how to design an α-helical protein. Proteins 7:1–15

Comery TA, Harris JB, Willems PJ, Oostra BA, Irwin SA, Weiler IJ, Greenough WT (1997a) Abnormal dendritic spines in fragile X knockout mice: maturation and pruning deficits. Proc Natl Acad Sci USA 94:5401–5404

Comery TA, Irwin SA, Patel ME, Gilbert ME, Kooy RF, Willems PJ, Oostra BA, Greenough WT (1997b) Abnormal dendritic spine morphology in fragile X syndrome: a role for the fragile-X mental retardation protein in spine maturation and pruning. Proc Soc Neurosci: 96.9

Coy JF, Sedlacek Z, Bächner D, Hameister H, Joos S, Lichter P, Delius H, Poustka A (1995) Highly conserved 3' UTR and expression pattern of FXR1 points to a divergent gene regulation of FXR1 and FMR1. Hum Mol Genet 4:2209–2218

Cronister A, Schreiner R, Wittenberger M, Amiri K, Harris K, Hagerman RJ (1991) The heterozygous fragile X female: historical, physical, cognitive and cytogenetic features. Am J Med Genet 38:269–274

Dahl N, Hu LJ, Chery M, Fardeau M, Gilgenkrantz S, Nivelon-Chevallier A, Sidaner-Noisette I, Mugneret F, Gouyon JB, Gal A (1995) Myotubular myopathy in a girl with a deletion at Xq27-q28 and unbalanced X inactivation assigns the MTM1 gene to a 600-kb region. Am J Hum Genet 56:1108–1115

David G, Abbas N, Stevanin G, Dürr A, Yvert G, Cancel G, Weber C, Imbert G, Saudou F, Antoniou E, Drabkin H, Gemmill R, Giunti P, Benomar A, Wood N, Ruberg M, Agid Y, Mandel J-L, Brice A (1997) Cloning of the SCA7 gene reveals a highly unstable CAG repeat expansion. Nat Genet 17:65–70

De Boulle K, Verkerk AJMH, Reyniers E, Vits L, Hendrickx J, van Roy B, van den Bos F, de Graaff E, Oostra BA, Willems PJ (1993) A point mutation in the FMR1 gene associated with fragile X mental retardation. Nat Genet 3:31–35

Deelen W, Bakker C, Halley DJ, Oostra BA (1994) Conservation of CGG region in FMR1 gene in mammals. Am J Med Genet 51:513–516

de Graaff E, Rouillard P, Willems PJ, Smits APT, Rousseau F, Oostra BA (1995a) Hotspot for deletions in the CGG repeat region of FMR1 in fragile X patients. Hum Mol Genet 4:45–49

de Graaff E, Willemsen R, Zhong N, de Die-Smulders CEM, Brown WT, Freling G, Oostra B (1995b) Instability of the CGG repeat and expression of the FMR1 protein in a male fragile X patient with a lung tumor. Am J Hum Genet 57:609–618

de Graaff E, De Vries BBA, Willemsen R, van Hemel JO, Mohkamsing S, Oostra BA, van den Ouweland AMW (1996) The fragile X phenotype in a mosaic male with a deletion showing expression of the FMR1 protein in 28 % of the cells. Am J Med Genet 64:302–308

Deissler H, Wilm M, Genç B, Schmitz B, Ternes T, Naumann F, Mann M, Doerfler W (1997) Rapid protein sequencing by tandem mass spectrometry and cDNA cloning of p20-CGGBP. J Biol Chem 272:16761–16768

de Vries BBA, Wiegers AM, Smits APT, Mohkamsing S, Duivenvoorden HJ, Fryns J-P, Curfs LMG, Halley DJJ, Oostra BA, van den Ouweland AMW, Niermijer MF (1996) Mental status of females with an FMR1 gene full mutation. Am J Hum Genet 58:1025–1032

de Vries LBA (1997) The fragile X syndrome. Clinical, genetic and large scale diagnostic studies among mentally retarded individuals. PhD Thesis, Erasmus University, Rotterdam

Devys D, Biancalana V, Rousseau F, Boué J, Mandel JL, Oberlé I (1992) Analysis of full fragile X mutations in foetal tissues and monozygotic twins indicate that abnormal methylation and somatic heterogeneity are established early in development. Am J Med Genet 43:208–216

Devys D, Lutz Y, Rouyer N, Bellocq J-P, Mandel J-L (1993) The FMR-1 protein is cytoplasmic, most abundant in neurons and appears normal in carriers of a fragile X premutation. Nat Genet 4:335–340

D'Hooge R, Nagels G, Franck F, Bakker CE, Reyniers E, Storm K, Kooy RF, Oostra BA, Willems PJ, De Deyn PP (1997) Mildly impaired water maze performance in male *Fmr1* knockout mice. Neuroscience 76:367–376

Eberhart DE, Warren ST (1996) Nuclease sensitivity of permeabilized cells confirms altered chromatin formation at the fragile X locus. Somat Cell Mol Genet 22:435–441

Eberhart DE, Malter HE, Feng Y, Warren ST (1996) The fragile X mental retardation protein is a ribonucleoprotein containing both nuclear localisation and nuclear export signals. Hum Mol Genet 5:1083–1091

Eichler EE, Richards S, Gibbs RA, Nelson DL (1993) Fine structure of the human FMR1 gene. Hum Mol Genet 2:1147–1153

Eichler EE, Holden JJA, Popovich BW, Reiss AL, Snow K, Thibodeau SN, Richards CS, Ward PA, Nelson DL (1995a) Length of uninterrupted CGG repeats determines instability in the *FMR1* gene. Nat Genet 8:88–94

Eichler EE, Kunst CB, Lugenbeel KA, Ryder OA, Davison D, Warren ST, Nelson DL (1995b) Evolution of the cryptic *FMR1* CGG repeat. Nat Genet 11:301–307

Falik-Zaccai TC, Shachak E, Yalon M, Lis Z, Borochowitz Z, Macpherson JN, Nelson DL, Eichler EE (1997) Predisposition to the fragile X syndrome in Jews of Tunisian descent is due to the absence of AGG interruptions on a rare mediterranean haplotype. Am J Hum Genet 60:103–112

Feng Y, Zhang F, Lokey LK, Chastain JL, Lakkis L, Eberhart D, Warren ST (1995) Translational suppression by trinucleotide repeat expansion at FMR1. Science 268:731–734

Feng Y, Lakkis L, Devys D, Warren ST (1995a) Quantitative comparison of FMR1 gene expression in normal and premutation alleles. Am J Hum Genet 56:106–113

Feng Y, Gutekunst C-A, Eberhart DE, Yi H, Warren ST, Hersch SM (1997b) Fragile X mental retardation protein: nucleocytoplasmic shuttling and association with somatodendritic ribosomes. J Neurosci 17:1539–1547

Fischer U, Huber J, Boelens WC, Mattaj IW, Lührmann R (1995) The HIV-1 Rev activation domain is a nuclear export signal that accesses an export pathway used by specific cellular RNAs. Cell 82:475–483

Fridell RA, Benson RE, Hua J, Bogerd HP, Cullen BR (1996) A nuclear role for the fragile X mental retardation protein. EMBO J 15:5408–5414

Fritz CC, Zapp ML, Green MR (1995) A human nucleoporin-like protein that specifically interacts with HIV Rev. Nature 376:530–533

Fryns J-P (1986) The female and the fragile X: a study of 144 obligate female carriers. Am J Med Genet 23:157–169

Fu Y-H, Kuhl DPA, Pizutti A, Pieretti M, Sutcliffe JS, Richards S, Verkerk AJMH, Holden JJA, Fenwick RG, Warren ST, Oostra BA, Nelson DL, Caskey CT (1991) Variation of the CGG repeat at the fragile X site results in genetic instability: resolution of the Sherman paradox. Cell 67:1047–1058

Gecz J, Gedeon AK, Sutherland GR, Mulley JC (1996) Identification of the gene FMR2, associated with FRAXE mental retardation. Nat Genet 13:105–108

Gecz J, Oostra BA, Hockey A, Carbonell P, Turner G, Haan EA, Sutherland GR, Mulley JC (1997) FMR2 expression in families with *FRAXE* mental retardation. Hum Mol Genet 6:435–441

Gedeon AK, Baker E, Robinson H, Partington MW, Gross B, Manca A, Korn B, Poustka A, Yu S, Sutherland GR, Mulley JC (1992) Fragile X syndrome without CCG amplification has an FMR1 deletion. Nature Genet 1:341–344

Gedeon AK, Keinänen M, Adès LC, Kääriäinen H, Gécz J, Baker E, Sutherland GR, Mulley JC (1995) Overlapping submicroscopic deletions in Xq28 in two unrelated boys with developmental disorders: identification of a gene near FRAXE. Am J Hum Genet 56:907–914

Gericke GS (1995) A paradigmatic shift in the approach to neuropsychiatric gene linkage may require an anthropogenetic perspective. Med Hypotheses 45:517–522

Giraud F, Ayme S, Mattei JF, Mattei MG (1976) Constitutional chromosomal breakage. Hum Genet 34:125–136

Gispert S, Santos N, Damen R, Voit T, Schulz J, Klockgether T, Orozco G, Kreuz F, Weissenbach J, Auburger G (1995) Autosomal dominant familial spastic paraplegia: reduction of the FSP1 candidate region on chromosome 14q to 7 cM and locus heterogeneity. Am J Hum Genet 56:183–187

Glover TW, Stein CK (1988) Chromosome breakage and recombination at fragile sites. Am J Hum Genet 43:265–273

Godde JS, Kass SU, Hirst MC, Wolffe AP (1996) Nucleosome assembly on methylated CGG triplet repeats in the fragile X mental retardation gene 1 promotor. J Biol Chem 271:24325–24328

Godfraind J-M, Reyniers E, De Boulle K, D'Hooge R, De Deyn PP, Bakker CE, Oostra BA, Kooy RF, Willems PJ (1996) Long-term potentiation in the hippocampus of fragile X knockout mice. Am J Med Genet 64:246–251

Görlich D, Mattaj IW (1996) Nucleocytoplasmic transport. Science 271:1513–1518

Gu Y, Nakamura T, Alder H, Prasad R, Canaani O, Cimino G, Croce CM, Canaani E (1992) The t(4;11) chromosome translocation of human acute leukemias fuses the ALL-1 gene, related to Drosophila trithorax, to the AF-4 gene. Cell 71:701–708

Gu YH, Lugenbeel KA, Vockley JG, Grody WW, Nelson DL (1994) A de novo deletion in FMR1 in a patient with developmental delay. Hum Mol Genet 3:1705-1706.

Gu Y, Shen Y, Gibbs RA, Nelson DL (1996) Identification of FMR2, a novel gene associated with the FRAXE CCG repeat and CpG island. Nat Genet 13:109–113

Hagerman RJ (1996) Physical and behavioral phenotype. In: Hagerman RJ, Cronister A (eds) Fragile X syndrome: diagnosis, treatment, and research. Johns Hopkins University Press, Baltimore, Massachusetts, pp 3–87

Hagerman RJ, Jackson K, Amiri K, Silverman AC, O'Connor R, Sobesky W (1992) Girls with fragile X syndrome: physical and neurocognitive status and outcome. Pediatrics 89:395–400

Hagerman RJ, Hull CE, Safanda JF, Carpenter I, Staley LW, O'Connor RA, Seydel C, Mazzocco MMM, Snow K, Thibodeau SN, Kuhl D, Nelson DL, Caskey CT, Taylor AK (1994) High functioning fragile X males: demonstration of an unmethylated fully expanded FMR-1 mutation associated with protein expression. Am J Med Genet 51:298–308

Hamel BCJ, Smits APT, de Graaff E, Smeets DFCM, Schoute F, Eussen BHJ, Knight SJL, Davies KE, Assman-Hulsmans CFCH, Oostra BA (1994) Segregation of FRAXE in a large family: clinical, psychometric, cytogenetic, and molecular data. Am J Hum Genet 55:923–931

Hansen RS, Canfield TK, Lamb MM, Gartler SM, Laird CD (1993) Association of fragile X syndrome with delayed replication of the FMR1 gene. Cell 73:1403–1409

Hansen RS, Canfield TK, Fjeld AD, Mumm S, Laird CD, Gartler SM (1997) A variable domain of delayed replication in FRAXA fragile X chromosomes: X inactivation-like spread of late replication. Proc Natl Acad Sci USA 94:4587–4592

Harris S, Moncrieff C, Johnson K (1996) Myotonic dystrophy: will the real gene please step forward! Hum Mol Genet 5:1417–1423

Hart PS, Olson SM, Crandall K, Tarleton J (1995) Fragile X syndrome resulting from a 400 bp deletion within the FMR1 gene. Am J Hum Genet 57:A1395

Hergersberg M, Matsuo K, Gassmann M, Schaffner W, Lüscher B, Rülicke T, Aguzzi A (1995) Tissue-specific expression of a FMR1/ß-galactosidase fusion gene in transgenic mice. Hum Mol Genet 4:359–366

Hewett DR, Handt O, Hobson L, Mangelsdorf M, Eyre H, Sutherland GR, Schuffenhauer S, Richards RI (1997) Positional cloning of the bromodeoxuridine-inducible fragile site on human chromosome 10q25.2. Am J Hum Genet 61:723

Hinds HL, Ashley CT, Sutcliffe JS, Nelson DL, Warren ST, Housman DE, Schalling M (1993) Tissue specific expression of FMR-1 provides evidence for a functional role in fragile X syndrome. Nat Genet 3:36–43

Hinton VJ, Brown WT, Wisniewski K, Rudelli RD (1991) Analysis of neocortex in three males with the fragile X syndrome. Am J Med Genet 41:289–294

Hirst M, Grewal P, Flannery A, Slatter R, Maher E, Barton D, Fryns JP, Davies K (1995) Two new cases of FMR1 deletion associated with mental impairment. Am J Hum Genet 56:67-74

Holden JJA, Julien-Inalsingh C, Chalifoux M, Wing M, Scott E, Fidler K, Swift I, Maidment B, Knight SJL, Davies KE, White BN (1996) Trinucleotide repeat expansion in the FRAXE locus is not common among institutionalized individuals with non-specific developmental disabilities. Am J Med Genet 64:420–423

Izakovic V (1984) Homozygosity for fragile site at 17p12 in a 28-year old healthy man. Hum Genet 68:340–341

Jacobsen P, Hauge M, Henningsen K, Hobolth N, Mikkelsen M, Philip J (1973) An (11;21) translocation in four generations with chromosome 11 abnormalities in the offspring. A clinical, cytogenetical, and gene marker study. Hum Hered 23:568–585

Jalal SM, Lindor NM, Michels VV, Buckley DD, Hoppe DA, Sarkar G, Dewald GW (1993) Absence of chromosome fragility at 19q13.3 in patients with myotonic dystrophy. Am J Med Genet 46:441–443

Jones C, Penny L, Mattina T, Yu S, Baker E, Voullaire L, Langdon WY, Sutherland GR, Richards RI, Tunnacliffe A (1995) Association of a chromosome deletion syndrome with a fragile site within the proto-oncogene CBL2. Nature 376:145–149

Kähkönen M, Tengström C, Alitalo T, Matilainen R, Kaski M, Airaksinen E (1989) Population cytogenetics of folate-sensitive fragile sites. II. Autosomal rare fragile sites. Hum Genet 82:3–8

Kambach C, Mattaj IW (1992) Intracellular distribution of the U1A protein depends on active transport and nuclear binding to U1 snRNA. J Cell Biol 118:11–21

Kawaguchi Y, Okamoto T, Taniwaki M, Aizawa M, Inoue M, Katayama S, Kawakami H, Nakamura S, Nishimura M, Akiguchi I, Kimura J, Narumiya S, Kakizuka A (1994) CAG expansions in a novel gene from Machado-Joseph disease at chromosome 14q32.1. Nat Genet 8:221–227

Kemper MB, Hagerman RJ, Ahmad RS, Mariner R (1988a) Cognitive profiles and the spectrum of clinical manifestations in heterozygous fragile (X) females. Am J Med Genet 23:139–156

Kemper MB, Hagerman RJ, Altshul-Stark D (1988b) Cognitive profiles of boys with the fragile X syndrome. Am J Med Genet 30:191–200

Khandjian EW, Fortin A, Thibodeau A, Tremblay S, Côté F, Devys D, Mandel J-L, Rousseau F (1995) A heterogeneous set of FMR1 proteins is widely distributed in mouse tissues and is modulated in cell culture. Hum Mol Genet 4:783–789

Khandjian EW, Corbin F, Woerly S, Rousseau F (1996) The fragile X mental retardation protein is associated with ribosomes. Nat Genet 12:91–93

Kiledjian M, Dreyfuss G (1992) Primary structure and binding activity of the hnRNP U protein: binding RNA through RGG box. EMBO J 11:2655–2664

Knight SJL, Flannery AV, Hirst MC, Campbell L, Christodoulou Z, Phelps SR, Pointon J, Middleton-Price HR, Barnicoat A, Pembrey ME, Holland J, Oostra BA, Bobrow M, Davies KE (1993) Trinucleotide repeat amplification and hypermethylation of a CpG island in FRAXE mental retardation. Cell 74:127–134

Knight SJL, Voelckel MA, Hirst MC, Flannery AV, Moncla A, Davies KE (1994) Triplet repeat expansion at the FRAXE locus and X-linked mild mental handicap. Am J Hum Genet 55:81–86

Koide R, Ikeuchi T, Onodera O, Tanaka H, Igarashi S, Endo K, Takahashi H, Kondo R, Ishikawa A, Hayashi T, Saito M, Tomoda A, Miike T, Naito H, Ikuta F, Tsuji S (1994) Unstable expansion of CAG repeat in hereditary dentatorubral-pallidoluysian atrophy (DRPLA). Nat Genet 6:9–13

Kooy RF, D'Hooge R, Reyniers E, Bakker CE, Nagels G, De Boulle K, Storm K, Clincke G, De Deyn PP, Oostra BA, Willems PJ (1996) Transgenic mouse model for the fragile X syndrome. Am J Med Genet 64:241–245

Kooy RF, Oostra BA, Willems PJ (1997) Molecular detection of dynamic mutations. In: Adolph KW (ed) Human genome methods. CRC Press, Boca Raton, Florida, pp 23–53

Kremer EJ, Pritchard M, Lynch M, Yu S, Holman K, Baker E, Warren ST, Schlessinger D, Sutherland GR, Richards RI (1991) Mapping of DNA instability at the fragile X to a trinucleotide repeat sequence p(CCG)n. Science 252:1711-1714

Kunst CB, Warren ST (1994) Cryptic and polar variation of the fragile X repeat could result in predisposing normal alleles. Cell 77:853–861

La Spada AR, Wilson EM, Lubahn DB, Harding AE, Fischbeck KH (1991) Androgen receptor gene mutations in X-linked spinar and bulbar muscular atrophy. Nature 352:77–79

Lavedan CN, Garrett L, Nussbaum RL (1997) Trinucleotide repeats $(CGG)_{22}TGG(CGG)_{43}TGG(CGG)_{21}$ from the fragile X gene remain stable in transgenic mice. Hum Genet 100:407–414

LeBeau MM, Rowley JD (1984) Heritable fragile sites and cancer. Nature 308:607–608

Loesch DZ, Hay DA (1988) Clinical features and reproductive patterns in fragile X female heterozygotes. J Med Genet 25:407–414

Lubs HA (1969) A marker X chromosome. Am J Hum Genet 21:231–244

Lugenbeel KA, Peier AM, Carson NL, Chudley AE, Nelson DL (1995) Intragenic loss of function mutations demonstrate the primary role of FMR1 in fragile X syndrome. Nat Genet 10:483–485

MacDonald ME, Ambrose CM, Duyao MP, Myers RH, Lin C, Srinidhi L, Barnes G, Taylor SA, James M, Groot N, MacFarlane H, Jenkins B, Anderson MA, Wexler NS, Gusella JF, Bates GP, Baxendale S, Hummerich H, Kirby S, North M, Youngman S, Mott R, Zehetner G, Sedlacek Z, Poustka A, Frischauf A-M, Lehrach H, Buckler AJ, Church D, Doucette-Stam L, O'Donovan MC, Riba-Ramirez L, Shah M, Stanton VP, Strobel SA, Draths KM, Wales JL, Dervan P, Housman DE, Altherr M, Shiang R, Thompson L, Fiedler T, Wasmuth JJ, Tagle D, Valdes J, Elmer L, Allard M, Castilla L, Swaroop M, Blanchard K, Collins FS, Snell R, Holloway T, Gillespie K, Shaw D, Harper PS (1993) A novel gene containing a trinucleotide repeat that is expanded and unstable on Huntington's disease chromosomes. Cell 72:971–983

Maes B, Fryns J-P, Van Walleghem M, Van den Berghe H (1994) Cognitive functioning and information processing of adult mentally retarded men with fragile-X syndrome. Am J Med Genet 50:190–200

Malmgren H, Steén-Bondeson M-L, Gustavson K-H, Seémanova E, Holmgren G, Oberlé I, Mandel J-L, Pettersson U, Dahl N (1992) Methylation and mutation patterns in the fragile X syndrome. Am J Med Genet 43:268–278

Malter HE, Iber JC, Willemsen R, de Graaff E, Tarleton JC, Leisti J, Warren ST, Oostra BA (1997) Characterization of the full fragile X syndrome mutation in foetal gametes. Nat Genet 15:165–169

Martin JP, Bell J (1943) A pedigree of mental defect showing sex-linkage. J Neurol Psychol 6:154–157

Mazzocco MMM, Pennington B, Hagerman RJ (1993) The neurocognitive phenotype of female carriers of fragile X: additional evidence for specificity. J Dev Behav Pediatr 14:328–335

McConcie-Rosell A, Lachiewicz AA, Spiridigliozzi GA, Tarleton J, Schoenwald S, Phelan MC, Goonewardena P, Ding X, Brown WT (1993) Evidence that methylation of the FMR1 locus is responsible for variant phenotypic expression of the fragile X syndrome. Am J Hum Genet 53:800–809

McInnis MG, McMahon FJ, Chase GA, Simpson SG, Ross CA, DePaulo JR Jr (1993) Anticipation in bipolar affective disorder. Am J Hum Genet 53:385–390

Meijer H, De Graaff E, Merckx DML, Jongbloed RJE, De Die-Smulders CEM, Engelen JJM, Fryns JP, Curfs PMG, Oostra BA (1994) A deletion of 1.6 kb proximal to the CGG repeat of the FMR1 gene causes the clinical phenotype of the fragile X syndrome. Hum Mol Genet 3:615-620.

Melberg A, Arnell H, Dahl N, Stålberg E, Raininko R, Oldfors A, Bakall B, Lundberg PO, Holme E (1996) Anticipation of autosomal dominant progressive external ophthalmoplegia with hypogonadism. Muscle Nerve 19:1561–1569

Merenstein SA, Sobesky WE, Taylor AK, Riddle JE, Tran HX, Hagerman RJ (1996) Molecular-clinical correlations in males with an expanded FMR1 mutation. Am J Med Genet 64:388–394

Meyer BE, Malim MH (1994) The HIV-1 Rev trans-activator shuttles between the nucleus and the cytoplasm. Genes Dev 8:1538–1547

Mila M, Castellvi, Barcelo A, Sanchez A, Jimenez D, Mandel J-L, Estivill X (1995) Mutations in the CpG island of the FMR1 gene: are they responsible for fragile X syndrome? Am J Hum Genet 57:A1273

Mila M, Castellvi-Bel S, Sanchez A, Lazaro C, Villa M, Estivill X (1996) Mosaicism for the fragile X syndrome full mutation and deletions within the CGG repeat of the FMR1 gene. J Med Genet 33:338-340

Moran TH, Reeves RH, Rogers D, Fisher E (1996) Ain't misbehavin' - it's genetic! Nat Genet 12:115-116

Mornet E, Bogyo A, Deluchat C, Simon-Boouy B, Mathieu M, Thepot F, Grisard M.-C, Leguern E, Boue J, Boue A (1993) Molecular analysis of a ring chromosome X in a family with fragile X syndrome. Hum Genet 92:373-378

Morris R (1984) Developments of a water-maze procedure for studying spatial learning in the rat. J Neurosci Methods 11:47-60

Morton JE, Bundey S, Webb TP, MacDonald F, Rindl PM, Bullock S (1997) Fragile X syndrome is less common than previously estimated. J Med Genet 34:1-5

Moutou C, Vincent M-C, Biancalana V, Mandel J-L (1997) Transition from premutation to full mutation in fragile X syndrome is likely to be prezygotic. Hum Mol Genet 6:971-979

Mulley JC, Yu S, Loesch DZ, Hay DA, Donnelly A, Gedeon AK, Carbonell P, López I, Glover G, Gabarrón I, Yu PWL, Baker E, Haan EA, Hockey A, Knight SJL, Davies KE, Richards RI, Sutherland GR (1995) FRAXE and mental retardation. J Med Genet 32:162-169

Murgia A, Polli R, Vinanzi C, Salis M, Drigo P, Artifoni L, Zacchello F (1996) Amplification of the Xq28 FRAXE repeats: extreme phenotype variability? Am J Med Genet 64:441-444

Murray A, Macpherson JN, Pound MC, Sharrock A, Youings SA, Dennis NR, McKechnie N, Linehan P, Morton NE, Jacobs PA (1997) The role of size, sequence and haplotype in the stability of FRAXA and FRAXE alleles during transmission. Hum Mol Genet 6:173-184

Musco G, Stier G, Joseph C, Castiglione Morelli MA, Nilges M, Gibson TJ, Pastore A (1996) Three-dimensional structure and stability of the KH domain: molecular insights into the fragile X syndrome. Cell 85:237-245

Nancarrow JK, Kremer E, Holman K, Eyre H, Doggett NA, Le Paslier D, Callen DF, Sutherland GR, Richards RI (1994) Implications of FRA16A structure for the mechanism of chromosomal fragile site genesis. Science 264:1938-1941

Nandabalan K, Price L, Roeder GS (1993) Mutations in U1 snRNA bypass the requirement for a cell type-specific RNA splicing factor. Cell 73:407-415

Nasir J, Floresco SB, O'Kusky JR, Diewert VM, Richman JM, Zeisler J, Borowski A, Marth JD, Phillips AG, Hayden MR (1995) Targeted disruption of the Huntington's disease gene results in embryonic lethality and behavioral and morphological changes in heterozygotes. Cell 81:811-823

Ohta M, Inoue H, Cotticelli MG, Kastury K, Baffa R, Palazzo J, Siprashvili Z, Mori M, McCue P, Druck T, Croce CM, Huebner K (1996) The FHIT gene, spanning the chromosome 3p14.2 fragile site and renal carcinoma-associated t(3;8) breakpoint, is abnormal in digestive tract cancers. Cell 84:587-597

Oostra BA, Willems PJ (1995) A fragile gene. BioEssays 17:941-947

Oostra BA, Willems PJ, Verkerk AJMH (1993a) Fragile X syndrome: a growing gene. In: Davies KE, Tilghman SM (eds) Genome maps and neurological disorders. Cold Spring Harbor Laboratory Press, New York, pp 45-75

Oostra BA, Jacky PB, Brown WT, Rousseau F (1993b) Guidelines for the diagnosis of fragile X syndrome. J Med Genet 30: 410-413

Orr HT, Chung M-y, Banfi S, Kwiatkowski TJ Jr, Servadio A, Beaudet AL, McCall AE, Duvick LA, Ranum LPW, Zoghbi HY (1993) Expansion of an unstable trinucleotide CAG repeat in spinocerebellar ataxia type 1. Nat Genet 4:221-226

Panté N, Aebi U (1996) Toward the molecular dissection of protein import into nuclei. Curr Opin Cell Biol 8:397-406

Parrish JE, Oostra BA, Verkerk AJMH, Richards CS, Reynolds J, Spikes AS, Shaffer LG, Nelson DL (1994) Isolation of a GCC repeat showing expansion in FRAXF, a fragile site distal to FRAXA and FRAXE. Nat Genet 8:229-235

Payami H, Bernard S, Larsen K, Kaye J, Nutt J (1995) Genetic anticipation in Parkinson's disease. Neurology 45:135-138

Penny LA, Dell'Aquila M, Jones MC, Bergoffen J, Cunniff C, Fryns J-P, Grace E, Graham Jr JM, Kousseff B, Mattina T, Syme J, Voullaire L, Zelante L, Zenger-Hain J, Jones OW, Evans GA (1995) Clinical and molecular characterization of patients with distal 11q deletions. Am J Hum Genet 56:676–683

Pieretti M, Zhang F, Fu Y-H, Warren ST, Oostra BA, Caskey CT, Nelson DL (1991) Absence of expression of the FMR-1 gene in fragile X syndrome. Cell 66:817–822

Pivnick EK, Velagaleti GVN, Wilroy RS, Smith ME, Rose SR, Tipton RE, Tharapel AT (1996) Jacobsen syndrome: report of a patient with severe eye anomalies, growth hormone deficiency, and hypothyroidism associated with deletion 11(q23q25) and review of 52 cases. J Med Genet 33:772–778

Price DK, Zhang F, Ashley CT Jr, Warren ST (1996) The chicken FMR1 gene is highly conserved with a CCT 5'-untranslated repeat and encodes an RNA-binding protein. Genomics 31:3–12

Pulkkinen L, Mannermaa A, Kajanoja E, Ryynananen M, Saarikoski S (1995) Deletion in the FMR1 gene in a fragile-X-male. Am J Hum Genet 57:A1300

Pulst S-M, Nechiporuk A, Nechiporuk T, Gispert S, Chen X-N, Lopes-Cendes I, Pearlman S, Starkman S, Orozco-Diaz G, Lunkes A, DeJong P, Rouleau GA, Auburger G, Korenberg JR, Figueroa C, Sahba S (1996) Moderate expansion of a normally biallelic trinucleotide repeat in spinocerebellar ataxia type 2. Nat Genet 14:269–276

Purpura DP (1974) Dendritic spine "dysgenesis" and mental retardation. Science 186:1126–1128

Quan F, Zonana J, Gunter K, Peterson KL, Magenis RE, Popovich BW (1995) An atypical case of fragile X syndrome caused by a deletion that includes the FMR1 gene. Am J Hum Genet 56:1042–1051

Reeves RH, Irving NG, Moran TH, Wohn A, Kitt C, Sisodia SS, Schmidt C, Bronson RT, Davisson MT (1995) A mouse model for Down syndrome exhibits learning and behaviour deficits. Nat Genet 11:177–184

Reiss AL, Abrams MT, Greenlaw R, Freund L, Denckla MB (1995) Neurodevelopmental effects of the FMR1 full mutation in humans. Nat Med 1:159–167

Reyniers E, Vits L, De Boulle K, van Roy B, van Velzen D, de Graaff E, Verkerk AJMH, Jorens HZJ, Darby JK, Oostra B, Willems PJ (1993) The full mutation in the FMR-1 gene of male fragile X patients is absent in their sperm. Nat Genet 4:143–146

Reyniers E, Martin J-J, Cras P, Van Marck E, Handig I, Jorens HZJ, Oostra BA, Kooy RF, Willems PJ (1998) Post-mortem examination of two fragile brothers with a full FMR1 mutation. Am J Med Genet (in press)

Richards RI, Holman K, Yu S, Sutherland GR (1993) Fragile X syndrome unstable element, p(CCG)n, and other simple tandem repeat sequences are binding sites for specific nuclear proteins. Hum Mol Genet 2:1429–1435

Richards RI, Sutherland GR (1992) Dynamic mutations: a new class of mutations causing human disease. Cell 70:709–712

Rousseau F, Heitz D, Biancalana V, Blumenfeld S, Kretz C, Boue J, Tommerup N, Van Der Hagen C, DeLozier-Blanchet C, Croquette MF, Gilgenkranz S, Jalbert P, Voelckel MA, Oberlé I, Mandel JL (1991) Direct diagnosis by DNA analysis of the fragile X syndrome of mental retardation. N Engl J Med 325: 1673–1681

Rousseau F, Heitz D, Tarleton J, MacPherson J, Malmgren H, Dahl N, Barnicoat A, Mathew C, Mornet E, Tejada I, Maddalena A, Spiegel R, Schinzel A, Marcos JAG, Schorderet DF, Schaap T, Maccioni L, Russo S, Jacobs PA, Schwartz C, Mandel JL (1994a) A multicenter study on genotype-phenotype correlations in the fragile X syndrome, using direct diagnosis with probe StB12.3: the first 2,253 cases. Am J Hum Genet 55:225–237

Rousseau F, Robb LJ, Rouilard P, Derkaloustian VM (1994b) No mental retardation in a man with 40 % abnormal methylation at the FMR-1 locus and transmission of sperm cell mutations as premutations. Hum Mol Genet 3:927–930

Sanberg PR, Johnson DA, Moran TH, Coyle JT (1984) Investigating locomotion abnormalities in animal-models of extrapyramidal disorders - a commentary. Physiol Psychol 12:48–50

Schinzel A, auf der Maur P, Moser H (1977) Partial deletion of long arm of chromosome 11[del(11)(q23)]: Jacobsen syndrome. Two new cases and review of the clinical findings. J Med Genet 14:438–444

Schmidt M, Certoma A, Du Sart D, Kalitsis P, Leversha M, Fowler K, Sheffield L, Jack I, Danks DM (1990) Unusual X chromosome inactivation in a mentally retarded girl with an interstitial deletion Xq27: implications for the fragile X syndrome. Hum Genet 84:347-352

Schmidt C, Henkel B, Pöschl E, Zorbas H, Puschke WG, Gloe TR, Müller PK (1992) Complete cDNA sequence of chicken vigilin, a novel protein with amplified and evolutionary conserved domains. Eur J Biochem 206:625-634

Sherman SL, Morton NE, Jacobs PA, Turner G (1984) The marker (X) chromosome: a cytogenetic and genetic analysis. Ann Hum Genet 48:21-37

Sherman SL, Jacobs PA, Morton NE, Froster-Iskenius U, Howard-Peebles PN, Nielsen KB, Partington NW, Sutherland GR, Turner G, Watson M (1985) Further segregation analysis of the fragile X syndrome with special reference to transmitting males. Hum Genet 69:3289-3299

Sijbers J, Scheunders P, Verhoye M, Van der Linden A, Van Dyck D, Raman E (1997) Watershed-based segmentation of 3D MR data for volume quantization. Magn Reson Imaging 15:679-688

Simko A, Hornstein L, Soukup S, Bagamery N (1989) Fragile X syndrome: recognition in young children. Pediatrics 83:547-552

Siomi H, Dreyfuss G (1995) A nuclear localisation domain in the hnRNP A1 protein. J Cell Biol 118:551-560

Siomi H, Siomi MC, Nussbaum RL, Dreyfuss G (1993) The protein product of the fragile X gene, FMR1, has characteristics of an RNA binding protein. Cell 74:291-298

Siomi H, Choi M, Siomi MC, Nussbaum RL, Dreyfuss G (1994) Essential role for KH domains in RNA binding: impaired RNA binding by a mutation in the KH domain of FMR1 that causes fragile X syndrome. Cell 77:33-39

Siomi MC, Siomi H, Sauer WH, Srinivasan S, Nussbaum RL, Dreyfuss G (1995) FXR1, an autosomal homolog of the fragile X mental retardation gene. EMBO J 14:2401-2408

Siomi MC, Zhang Y, Siomi H, Dreyfuss G (1996) Specific sequences in the fragile X syndrome protein FMR1 and the FXR proteins mediate their binding to 60S ribosomal subunits and the interactions among them. Mol Cell Biol 16:3825-3832

Sittler A, Deyvs D, Weber C, Mandel J-L (1996) Alternative splicing of exon 14 determines nuclear or cytoplasmatic localisation of Fmr1 protein isoforms. Hum Mol Genet 5:95-102

Slegtenhorst-Eegdeman KE, de Rooij DG, Verhoef-Post M, Ruiz A, Uilenbroek TJ, van de Kant HJG, Bakker CE, Oostra BA, Grootegoed JA, Themmen APN (1998) Macro-orchidism in *Fmr1* knockout mice is caused by increased Sertoli cell proliferation during normal testes development. Endocrinology 139:156-162

Smeets DFCM, Scheres JMJC, Hustinx TWJ (1985) Heritable fragility at 11q13 and 12q13. Clin Genet 28:145-150

Smeets HJM, Smits APT, Verheij CE, Theelen JPG, Willemsen R, van de Burgt I, Hoogeveen AT, Oosterwijk JC, Oostra BA (1995) Normal phenotype in two brothers with a full FMR1 mutation. Hum Mol Genet 4:2103-2108

Snow K, Tester DJ, Kruckeberg KE, Schaid DJ, Thibodeau SN (1994) Sequence analysis of the fragile X trinucleotide repeat: implications for the origin of the fragile X mutation. Hum Mol Genet 3: 1543-1551

Sobesky WE, Pennington BF, Porter D, Hull CE, Hagerman RJ (1994) Emotional and neurocognitive deficits in fragile X. Am J Med Genet 51:378-385

Subramanian PS, Nelson DL, Chinault AC (1996) Large domains of apparent delayed replication timing associated with triplet expansion at FRAXA and FRAXE. Am J Hum Genet 59:407-416

Sutherland GR (1977) Fragile sites on human chromosomes: demonstration of their dependence on the type of tissue culture medium. Science 197:265-266

Sutherland GR (1991) Chromosomal fragile sites. Genet Anal Tech Appl 8:161-166

Sutherland GR, Baker E (1992) Characterisation of a new rare fragile site easily confused with the fragile X. Hum Mol Genet 1:111-113

Sutherland GR, Hecht F (1985) Fragile sites on human chromosomes. Oxford University Press, New York

Tamanini F, Meijer N, Verheij C, Willems PJ, Galjaard H, Oostra BA, Hoogeveen AT (1996) FMRP is associated to the ribosomes via RNA. Hum Mol Genet 5:809-813

Tamanini F, Willemsen R, van Unen L, Bontekoe C, Galjaard H, Oostra BA, Hoogeveen AT (1997) Differential expression of FMR1, FXR1 and FXR2 proteins in human brain and testis. Hum Mol Genet 6:1315–1322

Tarleton J, Richie R, Schwartz C, Rao K, Aylsworth AS, Lachiewicz A (1993) An extensive de novo deletion removing FMR1 in a patient with mental retardation and the fragile X syndrome phenotype. Hum Mol Genet 2:1973–1974

Tharapel AT, Anderson KP, Simpson JL, Martens PR, Wilroy R Jr, Llerena J Jr, Schwartz CE (1993) Deletion (X)(q26.1-->q28) in a proband and her mother: molecular characterization and phenotypic-karyotypic deductions. Am J Hum Genet 52:463–471

Theobald TM, Hay DA, Judge C (1987) Individual variation and specific cognitive deficits in the fra(X) syndrome. Am J Med Genet 28:1–11

Trottier Y, Imbert G, Poustka A. Fryns JP, Mandel JL (1994) Male with typical fragile X phenotype is deleted for part of the FMR1 gene and for about 100 kb of upstream region. Am J Med Genet 51:454–457.

Turecki G, Smith MdAC, Mari JJ (1995) Type I bipolar disorder associated with a fragile site on chromosome 1. Am J Med Genet 60:179–182

Turner G, Webb T, Wake S, Robinson H (1996) Prevalence of fragile X syndrome. Am J Med Genet 64:196–197

Verheij C, Bakker CE, de Graaff E, Keulemans J, Willemsen R, Verkerk AJMH, Galjaard H, Reuser AJJ, Hoogeveen AT, Oostra BA (1993) Characterization and localisation of the FMR-1 gene product associated with fragile X syndrome. Nature 363:722–724

Verheij C, de Graaff E, Bakker CE, Willemsen R, Willems PJ, Meijer N, Galjaard H, Reuser AJJ, Oostra BA, Hoogeveen AT (1995) Characterization of FMR1 proteins isolated from different tissues. Hum Mol Genet 4:895–901

Verkerk AJMH, Pieretti M, Sutcliffe JS, Fu Y-H, Kuhl DPA, Pizzuti A, Reiner O, Richards S, Victoria MF, Zhang F, Eussen BE, van Ommen G-JB, Blonden LAJ, Riggins GJ, Chastain JL, Kunst CB, Galjaard H, Caskey CT, Nelson DL, Oostra BA, Warren ST (1991) Identification of a gene (FMR-1) containing a CGG repeat coincident with a breakpoint cluster region exhibiting length variation in fragile X syndrome. Cell 65:905–914

Verkerk AJMH, de Graaff E, De Boulle K, Eichler EE, Konecki DS, Reyniers E, Manca A, Poustka A, Willems PJ, Nelson DL, Oostra BA (1993) Alternative splicing in the fragile X gene FMR1. Hum Mol Genet 2:399–404

Verma IC, Elango R (1994) Variable expression of clinical features of Martin Bell syndrome in younger patients. Indian Pediatr 31:433–438

Voullaire LE, Webb GC, Leversha MA (1987) Chromosome deletion at 11q23 in an abnormal child from a family with inherited fragility at 11q23. Hum Genet 76:202–204

Wang Y-H, Griffith J (1996a) Methylation of expanded CCG triplet repeat DNA from fragile X syndrome patients enhances nucleosome exclusion. J Biol Chem 271:22937–22940

Wang Y-H, Griffith JD (1996b) The $[(G/C)_3NN]_n$ motif: a common DNA repeat that excludes nucleosomes. Proc Natl Acad Sci USA 93:8863–8867

Wang Y-H, Amirhaeri S, Kang S, Wells RD, Griffith JD (1994) Preferential nucleosome assembly at the DNA triplet repeats from the myotonic dystrophy gene. Science 265:669–671

Wang Y-H, Gellibolian R, Shimizu M, Wells RD, Griffith J (1996) Long CCG triplet repeat blocks exclude nucleosomes: a possible mechanism for the nature of fragile sites in chromosomes. J Mol Biol 263:511–516

Weiler IJ, Irwin SA, Klintsova AY, Spencer CM, Brazelton AD, Miyashiro K, Comery TA, Patel B, Eberwine J, Greenough WT (1997) Fragile X mental retardation protein is translated near synapses in response to neurotransmitter activation. Proc Natl Acad Sci USA 94:5395–5400

Wells RD (1996) Molecular basis of genetic instability of triplet repeats. J Biol Chem 271: 2875–2878

Wen W, Meinkoth JL, Tsien RY, Taylor SS (1995) Identification of a signal for rapid export of proteins from the nucleus. Cell 82:463–473

Willems PJ (1994) Dynamic mutations hit double figures. Nat Genet 8:213–215

Willems PJ, van Roy B, De Boulle K, Vits L, Reyniers E, Beck O, Dumon JE, Verkerk AJMH, Oostra B (1992) Segregation of the fragile X mutation from an affected male to his normal daughter. Hum Mol Genet 1:511–515

Willems PJ, Reyniers E, Oostra BA (1995) An animal model for fragile X syndrome. Ment Retard Dev Dis Res Rev 1:298–302

Willemsen R, Mohkamsing S, De Vries B, Devys D, van den Ouweland A, Mandel JL, Galjaard H, Oostra B (1995) Rapid antibody test for fragile X syndrome. Lancet 345:1147–1148

Willemsen R, Bontekoe C, Tamanini F, Galjaard H, Hoogeveen A, Oostra BA (1996) Association of FMRP with ribosomal precursor particles in the nucleolus. Biochem Biophys Res Commun 225:27–33

Wöhrle D, Hennig I, Vogel W, Steinbach P (1993) Mitotic stability of fragile X mutations in differentiated cells indicates early post-conceptional trinucleotide repeat expansion. Nat Genet 4:140-142

Wolff DJ, Conroy J, Zurcher V, Gustashaw K, Ko L, VanDyke DL, Weiss L, Willard HF, Schwartz S (1995) Deletions of ~12 Mb of Xq including FMR1 results in a severe phenotype in a male and variable phenotypes in females depending upon the X inactivation pattern. Am J Hum Genet 57:A734

Yano-Yanagisawa H, Li Y, Wang H, Kohwi Y (1995) Single-stranded DNA binding proteins isolated from mouse brain recognize specific trinucleotide repeat sequences in vitro. Nucleic Acids Res 23:2654–2660

Yu S, Mangelsdorf M, Hewett D, Hobson L, Baker E, Eyre HJ, Lapsys N, Le Paslier D, Doggett NA, Sutherland GR, Richards RI (1997) Human chromosomal fragile site FRA16B is an amplified AT-rich minisatellite repeat. Cell 88:367–374

Yunis JJ, Soreng AL (1984) Constitutive fragile sites and cancer. Science 226:1199–1204

Zhang Y, O'Connor JP, Siomi MC, Srinivasan S, Dutra A, Nussbaum RL, Dreyfuss G (1995) The fragile X mental retardation syndrome protein interacts with novel homologs FXR1 and FXR2. EMBO J 14:5358–5366

Zhong N, Ju W, Curley D, Wang D, Pietrofesa J, Wu G, Shen Y, Pang C, Poon P, Liu X, Gou S, Kajanoja E, Ryynänen M, Dobkin C, Brown WT (1996) A survey of FRAXE allele sizes in three populations. Am J Med Genet 64:415–419

Zhuchenko O, Bailey J, Bonnen P, Ashizawa T, Stockton DW, Amos C, Dobyns WB, Subramony SH, Zoghbi HY, Chi Lee C (1997) Autosomal dominant cerebellar ataxia (SCA6) associated with small polyglutamine expansions in the a_{1A}-voltage-dependent calcium channel. Nat Genet 15:62–69

Molecular Genetics of Huntington's Disease

Marcy E. MacDonald[1]

1
Introduction

In Huntington's disease (HD), with its perplexing and paradoxical features of genetic inheritance, the use of molecular genetics has been championed for the identification of human disease genes. In turn, the discovery of the *HD* mutation, an expanded, unstable CAG repeat in a novel 4p16.3 gene, has solved many of the disorder's genealogical puzzles, provided a new tool for molecular testing and has spurred more precise descriptions of the clinical and pathological features of the disease. Moreover, knowledge of the defect has raised the possibility that HD shares aspects of a common pathogenic mechanism with a number of other CAG repeat disorders, making it likely that the development of an effective treatment for this inevitably fatal affliction will not occur in isolation.

2
Features of the Disease

George Huntington described the disease that today bears his name more than 100 years ago (Huntington 1872). As a third generation Long Island physician, Huntington was in a unique position to remark on the genetic features and clinical variability of the disorder, which he termed an "hereditary chorea" because of its flagrant, uncontrolled "dance-like" movements. Males and females were equally afflicted, and children inherited the disease from mothers and fathers without skipping generations. The harrowing symptoms manifested most often in midlife, progressing over a period of years to an inevitable death. Huntingtin likened the disease to "an heirloom from generations away back in the dim past", and later it was attributed to the inheritance of an autosomal dominant gene defect (Osler 1908). Today, the disorder is called Huntington's disease (HD) in recognition of its other features, the psychiatric and cognitive deficits that Huntington, mistakenly but in the fashion of the day, labeled insanity. Indeed, its unique symptoms are due to a selective neurodegeneration that ultimately claims its victims.

[1] Molecular Neurogenetics Unit, Massachusetts General Hospital East, Charlestown, Massachusetts 02129, USA

HD strikes people of all races, nationalities, and ethnic backgrounds (Vessie 1932; Harper 1992). In all cases the disease is caused by a single gene defect, an unstable expanded CAG repeat, which also explains the rare appearance of "sporadic" new mutations. In hindsight, molecular geneticists familiar with the non-Mendelian characteristics of trinucleotide repeat expansions could have anticipated the genetic defect from a close reading of Huntington's original report.

2.1
Clinical Features

HD is a progressive neurodegenerative disorder that can manifest at any age, although onset is usually in midlife (approximately 48 years). It is characterized by peculiar uncontrolled, dance-like movements (chorea), cognitive impairment and emotional decline (Martin and Gusella 1986; Folstein 1989). Symptoms begin insidiously, and the transition from slight motor restlessness to frank incessant chorea, sudden unintended movements of any part of the body, is inexorable. Patients, ultimately, are emaciated and incapacitated by choreiform movements that are unremitting except in sleep. Death follows 10–20 years after onset as they succumb to heart disease, aspiration pneumonia, or other complications of the uncontrolled movements. About 10 % of patients have onset before 20 years of age, and in these patients the disease follows a grim course that includes seizures and motor rigidity, myoclonus and dystonia rather than chorea (Folstein 1989; Merritt et al. 1969).

The neurologic symptoms may be accompanied or preceded by psychological and cognitive changes, compromising job performance and family life. Irritability, restlessness, erratic behavior, memory loss, and decreasing mental capacity are evident. Often increasing apathy and deepening bouts of depression prompt suicidal episodes. Clinical management of HD is palliative. Pharmacologic interventions alleviate motor symptoms and high caloric intake balances weight loss for a time but the disease cannot be prevented, nor is there a treatment that will delay its relentless advance.

2.2
Neuropathology

The symptoms are caused by a loss of neurons in the brain that occurs about the time that disease first becomes manifest. The basal ganglia and cortex are ravaged, a process that can be followed by magnetic resonance imaging (MRI) and positron emission tomography (PET). In the caudate nucleus, populations of enkephalin and substance P containing medium-sized spiny GABAergic projection neurons are the first to be affected, exhibiting wilted, recurved dendritic endings and changes in the density, shape and size of spines (Graveland et al. 1987). The large acetylcholine rich or smaller somatostatin and neuropeptide

Y containing aspiny interneurons are spared by the disease process (Ferrante et al. 1985; Albin et al. 1992). Indeed, the characteristic pattern of neuronal cell loss in the basal ganglia forms the basis for the neuropathological grading of HD (Vonsattel et al. 1985).

Neurons are also lost in deep layers V and VI of the cerebral cortex (de la Monte et al. 1988; Hedreen et al. 1991), and brain weight may be reduced by as much as 30 % in end stage disease. Other brain regions, including the thalamic and subthalamic nuclei, brain stem, and spinal cord, are less consistently affected and changes in these areas are likely to be secondary to the primary disease process.

2.3
Genetics

HD is caused by a highly penetrant inherited genetic defect that is transmitted in an autosomal dominant fashion. It has a prevalence of approximately 1 in 10,000 in Caucasian populations, but is found around the world occurring in other races at lower frequency (Harper 1992). Approximately 2.5 times more individuals are at risk for the disorder because of the midlife peak in age at onset and about 40 % of those at risk actually have the gene defect and are too young to exhibit symptoms of the disease (Fig. 1). HD affects males and females equally and can be transmitted from mothers and fathers, although the juvenile form of the disease tends to be inherited from fathers (Bird et al. 1974; Merritt et al. 1969). Most HD is inherited, but there are occurrences of spontaneous HD, in which an affected individual has transmitted the disease but has no previous family history (Bozza et al. 1995; Davis et al. 1994; Durr et al. 1995; Goldberg et al. 1993; Myers et al. 1993).

3
HD locus

The clinical and genetic characteristics of HD, including its distinct symptoms, midlife onset, unambiguous mode of inheritance, high penetrance and prevalence in the general population, made it an ideal disease to attack by genetic linkage (Gusella et al. 1984). In 1983, the disease gene, *HD*, was mapped to the vicinity of *D4S10*, an anonymous polymorphic DNA marker, by analysis of the inheritance of restriction fragment length polymorphisms in two large HD kindred of American and Venezuelan decent (Gusella et al. 1983). Genetic (Gilliam et al. 1987) and physical methods (Landegent et al. 1986; Gusella et al. 1985) subsequently assigned *HD* to the tip of the short arm of chromosome 4 in 4p16.3.

Linkage to *D4S10* demonstrated that most, if not all, cases of HD were likely to arise from defects in the same gene (Conneally et al. 1989). Moreover, *D4S10* was shown to be approximately 4 centimorgans (about 4 % recombination) from *HD*, providing the baseline of information needed to establish a DNA test

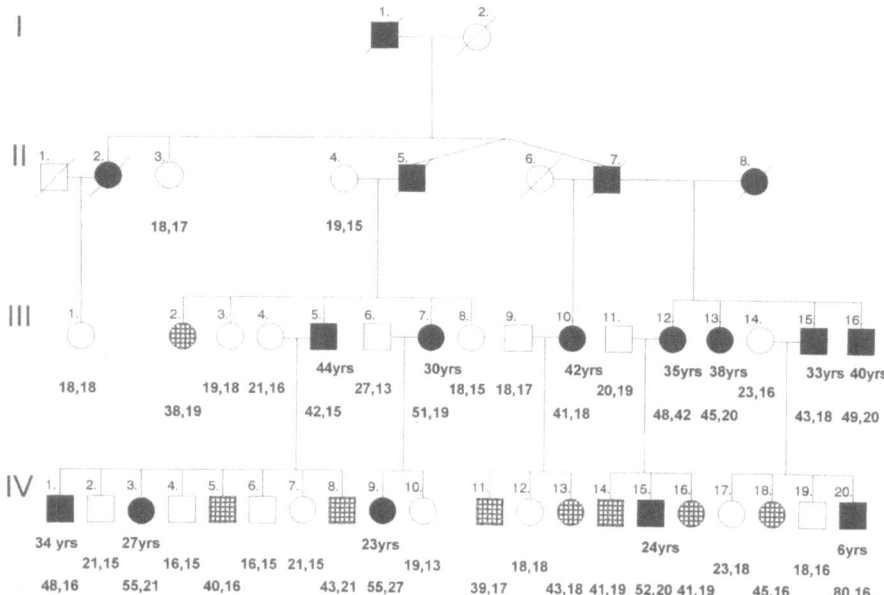

Fig. 1. Inheritance of HD and the *HD* mutation. An idealized four generation (I–IV) HD pedigree is shown with living and deceased (*slashed symbols*) males (*circles*) and females (*squares*) numbered sequentially. The age at onset of symptoms (in years) is given below for clinically diagnosed cases (*filled symbols*). HD is transmitted in a dominant fashion and, although approximately half the children of an affected parent are expected to develop disease symptoms many individuals in generation IV are at risk for the disorder as they have inherited the defect (*hatched symbols*) but are too young to display symptoms. The length of the *HD* CAG repeat at both alleles determined by direct DNA testing in symptomatic and presymptomatic individuals is given under each symbol. A comparison of the clinical and molecular genotyping data reveals a number of features of the *HD* mutation. The length of the expanded CAG repeat is inversely correlated with age at onset of symptoms, and alleles in the expanded (37–80 units) but not the normal (15–27) range change in size when transmitted from parent to child, exhibiting a tendency to increase in size, so that the longest alleles occur in generation IV. Striking increases in size occur upon transmission through the male germline and, in individual IV-20, a disease allele with 80 CAG repeats causes juvenile onset disease. By contrast, III-2, an elderly individual with affected siblings, has inherited the same disease chromosome with an expanded CAG repeat of 38 units, but does not have HD, illustrating the reduced penetrance of alleles in the 36- to 39- repeat range. The children of II-7 and II-8 can inherit the disease from either affected parent, and III-13, III-15 and III-16 are all typical HD heterozygotes, while III-12 has inherited the mutation from both parents and is an HD homozygote. The onset of clinical symptoms in the latter does not differ from that observed in her three heterozygous siblings, arguing that one copy of the expanded CAG repeat saturates the pathogenic mechanism

for asymptomatic, at-risk individuals with family members willing to participate in a genetic linkage test. Linkage to *D4S10* also revealed the complete phenotypic dominance of the disorder (Wexler et al. 1987; Myers et al. 1989) as affected HD homozygotes, with two copies of the defect, were indistinguishable clinically from their heterozygous siblings who have only one copy of the mutation.

The discovery of the *HD* defect was the culmination of the molecular genetic search that began with linkage to *D4S10* (Gusella and MacDonald 1995). Genetic and physical mapping studies by a consortium of investigators confined *HD* to a region of approximately 2 million bp of 4p16.3. Polymorphic DNA markers were developed, genetic and physical maps of the region constructed, and overlapping clone sets generated while recombination analysis in disease pedigrees and linkage disequilibrium between the disorder and genetic markers targeted the search area.

Ultimately, the location of the gene (Fig. 2) emerged from an analysis of haplotypes of multiallele markers, revealing that, although approximately two thirds of disease chromosomes had unrelated haplotypes and likely had independent orgins, a subset (approximately one third) shared a small (275 kb) region between *D4S95* and *D4S180* because they descended from a common ancestral chromosome (MacDonald et al. 1992). Highly polymorphic markers and Δ2642, a two allele polymorphism in linkage disequilibrium with HD, highlighted a segment of approximately 150 kb on these "major" HD haplotype chromosomes as the most likely location of the gene defect. Finally, analysis of candidate genes from this subregion merged with the search for genetic markers to reveal a polymorphic CAG repeat in the 5' end of interesting transcript 15 (IT15), which was expanded and unstable on disease chromosomes (Huntington's Disease Collaborative Research Group 1993). IT15 was assigned the locus name *HD*, signaling the start of the quest for the disease geneõs pathogenic mechanism.

4
HD Mutation: An Unstable Expanded CAG Repeat

The pure array of CAG repeats that is expanded on disease chromosomes is located near the 5õ end of a novel 4p16.3 gene. It is immediately adjacent to a broken array of CAG/CCG codons that contains a polymorphic (6 to 12 repeats), stably transmitted stretch of CCG triplets. Most normal chromosomes and the majority (> 90 %) of disease chromosomes possess seven CCG repeats (Rubinsztein et al. 1993; Andrew et al. 1994a). By contrast, the CAG repeat is highly polymorphic (Fig. 3). Normal chromosomes possess 6–34 CAG repeats that are inherited in a Mendelian fashion, whereas HD chromosomes have 36 to approximately 120 units that are inherited in a strikingly non-Mendelian manner (Huntington's Disease Collaborative Research Group 1993; Barron et al. 1993; De Rooij et al. 1993; Dode et al. 1993; Duyao et al. 1993; Kremer et al. 1993; 1994; Lucotte et al. 1994; Masuda et al. 1995; Norremolle et al. 1993; Novelletto et al. 1994; Persichetti et al. 1994; Snell et al. 1993; Soong and Wang 1995; Whitefield et al. 1996; Zuhlke et al. 1993b). Rare alleles with 36–39 repeats are found in exceptional individuals with HD or in the unaffected elderly relatives of patients with sporadic de novo cases of the disease, arguing that alleles in this range exhibit reduced penetrance (Rubinsztein et al. 1996; McNeil et al. 1997).

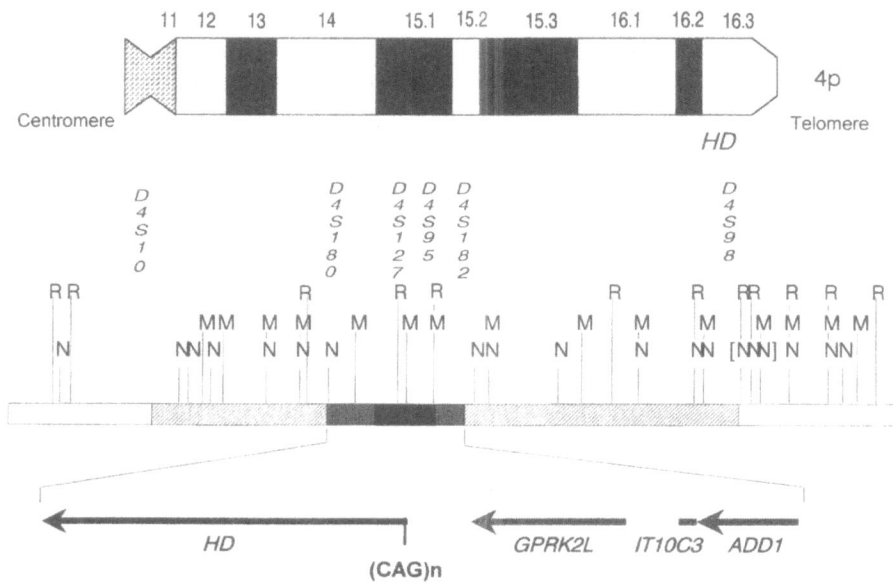

Fig. 2. Isolation of the HD gene by location cloning. The short arm of chromosome 4 (4p) is depicted with numbered (*above*) cytogenetic bands to indicate the location of the *HD* defect (*HD*) in 4p16.3 as determined by linkage analysis and physical mapping. *Below* A schematic long range restriction map of the HD region (approximately 3×10^6 bp of 4p16.3). *R* NruI site; *M* MluI site; *N* NotI; *brackets* denote sites that are partially cut. The progressive narrowing of the region is denoted by the *unshaded* region, which was eliminated by recombination events in HD families, highlighting the approximately 2 million bp region between *D4S10* and *D4S98* (*light shading*). Linkage disequilibrium and haplotype studies refined the defect to the approximately 500 kb region between *D4S180* and *D4S182* (*darker shading*) and further haplotype analysis placed the defect within the approximately 150 kb region between Δ2642 and *D4S127* (*darkest shading*). The segment between *D4S180* and *D4S182* is expanded below, depicting the location of the CAG repeat (CAG)$_n$ at the 5' end of the *HD* gene (*HD*) and neighboring candidate genes α-adducin (*ADD1*), a novel protein with homology to a tetracycline transporter (*IT10C3*), and a G-coupled protein receptor kinase (*GPRK2L*)

4.1
Instability

The length of the expanded HD CAG repeat changes, almost invariably increasing or decreasing in size, when transmitted from parent to child (Huntington's Collaborative Research Group 1993; Duyao et al. 1993; Norremolle et al. 1995; Snell et al. 1993; Trottier et al. 1994; Zuhlke et al. 1993; Gusella and MacDonald 1996; Fig. 4). In most cases, whether inherited from a mother or father, the length alterations are modest (less than six units), with a bias toward increases, but transmission through the male germline can result in large increases in size, even doubling the number of CAG repeats (De Rooij et al. 1993; Duyao et al. 1993; Goldberg et al. 1993; Norremolle et al. 1995a; Snell et al. 1993; Telenius et

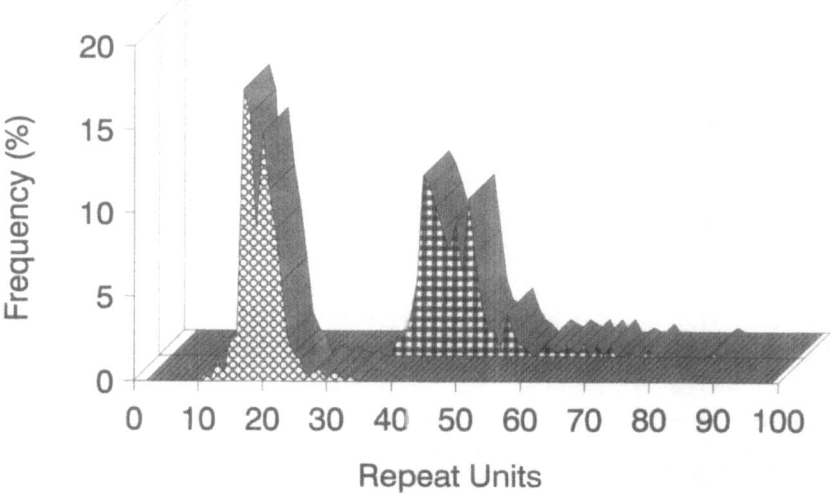

Fig. 3. Distribution of CAG repeat lengths on HD and normal chromosomes. The frequency of CAG allele sizes for normal (*dotted pattern*) and disease (*plaid pattern*) chromosomes, shown as a percentage of the total, is plotted against the number of CAG repeat units (Duyao et al. 1993). These chromosomes were drawn from individuals of disparate genetic and ethnic backgrounds from around the world. Rare chromosomes with allele sizes of 35 and 36 units that have been identified in families with a member who represents a new mutation to HD (HDCRG 1993; Myers 1993; Goldberg et al. 1993; Bozza et al. 1995; Davis et al. 1994; Durr et al. 1995) are not included in this data set

al. 1994; Trottier et al. 1994; Zuhlke et al. 1993). The latter occurs during spermatogenesis and can be detected by variation in the length of the CAG repeat in sperm DNA (MacDonald et al. 1993a; Leeflang et al. 1995). Expanded repeats of any size can change in length, but in sperm DNA the tendency of the allele to exhibit a dramatic spread in alleles of increasing size is enhanced as the length of the repeat increases above approximately 45 units. Pairs of monozygotic HD twins have identical expanded CAG repeat lengths, and similar clinical findings, suggesting that the intergenerational instability arises not in embryonic development, but rather during gametogenesis (MacDonald et al. 1993). Because the expanded repeat changes when transmitted through the maternal germline, it is likely that instability, albeit more restricted in magnitude, also occurs during oogenesis, but this has not been tested.

Occasional expansion of relatively rare CAG repeats of 34–36 units to lengths of 37 or more upon transmission through the male germline gives rise to a new disease causing allele and is the explanation for cases of sporadic HD (Bozza et al. 1995; Davis et al. 1994; Durr et al. 1995; Goldberg et al. 1993; Myers et al. 1993). Most of these events occur on the same major 4p16.3 haplotype that is found in approximately one third of unrelated HD families (Myers et al. 1993; Goldberg et al. 1993). It is likely, therefore, that these disease families share a common ancestor who was not affected by the disorder and that these major

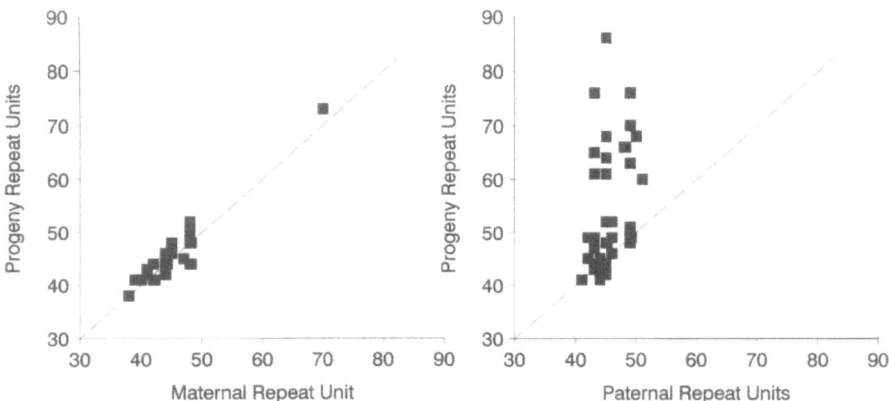

Fig. 4. Intergenerational instability of *HD* CAG repeat length. The length of the expanded CAG repeat (in units) on disease chromosomes in mothers (*left*) and fathers (*right*) is plotted against the length of the expanded repeat in corresponding children. The diagonal (*dotted line*) denotes no change in repeat length. The repeat length has changed, increasing or decreasing by a few units in the majority of the 25 maternal and 37 paternal transmissions, but in approximately one third of the cases transmission from fathers has produced a large increase in the size of the repeat. (Duyao et al. 1993)

haplotype bearing chromosomes today serve as a reservoir that can give rise to new sporadic HD cases. The frequency of these chromosomes in the general population is not known but must be quite low given that they represent less than 4 % of chromosomes in a population selected because a family member had HD. Our understanding of the mechanism that generates CAG repeat instability will increase as the effects of surrounding DNA sequences, disease status, parental age, and a host of other biological factors that have yet to be defined are investigated.

Somatic mosaicism is not a prominent feature of HD and does not explain its pathology as the expanded CAG repeat does not undergo significant instability in cell culture or in tissue until repeat size exceeds approximately 60 units (De Rooij et al. 1995; Zuhlke et al. 1993; Telenius et al. 1994). An intriguing and as yet unexplained observation is that, for the largest repeats which exhibit evidence of mosaicism in the cerebral cortex, striatum, blood, and other tissues, a different shorter CAG repeat is found in the cerebellum (Telenius et al. 1994). While it is not involved in the pathogenic mechanism, it is likely that this odd pattern of instability in the cerebellum serves as a lineage marker for the cells which ultimately form the adult tissue.

4.2
Clinical Correlates

The expanded HD CAG repeat is the only mutation resulting in the clinical and neuropathological symptoms of HD. All proven cases of familial (Andrew et al. 1994; Kremer et al. 1994) and sporadic HD (Andrew et al. 1994; Bozza et al. 1995; Davis et al. 1994; Durr et al. 1995; Goldberg et al. 1993; Myers et al. 1993) and clinically diagnosed, neuropathologically confirmed cases (Persichetti et al. 1995; Xuereb et al. 1996) possess an expanded CAG repeat. Much of the variation in age at onset of neurological and psychiatric symptoms and in severity of the disorder is explained by a strong inverse correlation with the number of CAG repeats in the expanded array (Fig. 5; Andrew et al. 1993; Duyao et al. 1993; Claes et al. 1995; Craufurd and Dodge 1993; Kremer et al. 1993; Stine et al. 1993; Whitefield et al. 1996; Gusella and MacDonald 1995; Gusella et al. 1997; Persichetti et al. 1995). Most HD patients have neurologic onset in midlife (35–55 years) due to repeats of 40–50 CAG units. The relatively rare juvenile form of the disease occurs in individuals with repeat lengths of more than 65 CAG units

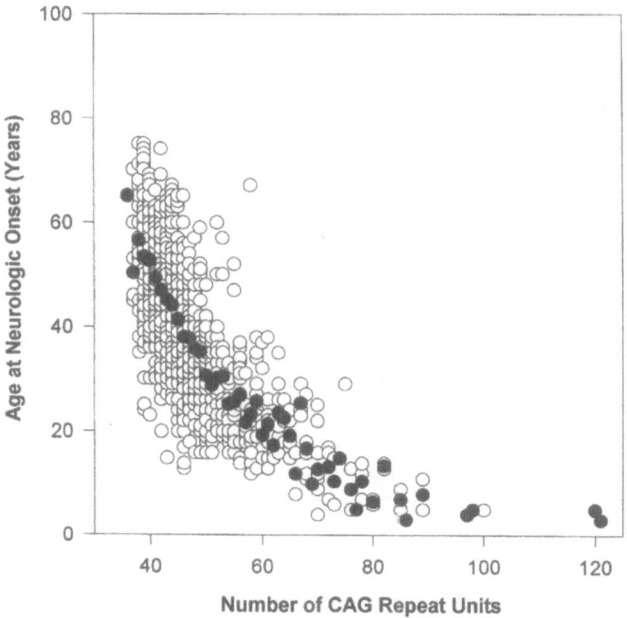

Fig. 5. Relationship of CAG repeat length and age at onset. The length of the CAG repeat in 1226 HD patients is plotted against the age at onset of neurologic disease symptoms. The mean age at onset (*open circles*) associated with any given CAG repeat length is also depicted (*filled circles*). In this data, set the correlation between the mean age at onset and CAG repeat length is highly significant across all HD allele sizes when fitted with an exponential decay model (r -0.97, P<0.00001; Gusella and MacDonald 1996) but the wide range of possible ages of onset associated with any given repeat length precludes an accurate estimate of when the disease will strike any single individual

that arise from shorter alleles by transmission through the male germline, thereby explaining the preponderance of paternal inheritance in these cases.

The length of the CAG expansion is the primary determinant of age at onset of clinical symptoms, but is not predictive of age at onset in any given HD patient (Gusella and MacDonald 1995; Gusella et al. 1997). The broad range of onset ages associated with each expanded allele points to the influence of unknown genetic, environmental, or stochastic modifying factors. This variation is most intriguing in the 36- to 39-CAG range, as these alleles typically cause disease late in life but exceptional individuals live to a very old age without manifestations of HD, suggesting that modifiers can determine whether the expanded allele is penetrant (McNeil et al. 1997; Rubinsztein et al. 1996). The identification of these factors, therefore, offers a promising route to finding agents that may delay onset of disease in HD patients with longer alleles.

The inverse correlation between CAG repeat length and age at onset of neurologic symptoms is best approximated by an exponential decay model in which the contribution of each additional CAG unit to reducing age at onset becomes progressively smaller as the size of the CAG expansion increases (Fig. 5; Gusella and MacDonald 1997). The length of the CAG repeat, therefore, may determine the rate of neuronal dysfunction or loss, and onset of disease may occur when the surviving functional capacity of some key population of neurons is diminished below a critical threshold. Extrapolation of cell counts in postmortem HD brain suggests that disease could become manifest when approximately 30 % of the neurons in the caudate nucleus are lost (Furtado et al. 1996). If this process occurs throughout the lifetime of an HD individual, some level of neuronal cell dysfunction or loss would be expected before onset of clinical symptoms and should be evident from close examination of brains from asymptomatic HD individuals. Alternatively, gradual neuronal cell dysfunction or loss does not occur, but rather the exponential relationship reflects the age at which CAG repeat length first triggers the pathogenic process.

CAG repeat length is also correlated with age at onset of pyschiatric symptoms and a similar statistical relationship to that seen with age at onset of neurologic disease suggests that the underlying pathologic process is similar or identical to that causing motor impairment (Persichetti et al. 1995). Similarly, age at death is strongly correlated with CAG repeat length, demonstrating that the size of the expanded allele is the primary determinant of the reduced life span in HD. However, the size of the expanded allele is not correlated with length of time from disease onset to death (Gusella et al. 1997), which is approximately 15 years regardless of CAG length. Indeed, the influence of CAG repeat length on measures of functional decline in HD patients is a subject of debate (Brandt et al. 1996; Kieburtz et al. 1994; Illarioshkin et al. 1994) and it is likely that this lack of consensus reflects the absence of a strong correlation. The process that determines disease progression, therefore, probably involves factors beyond the CAG length dependent neuronal cell death that determines onset of neurologic disease.

4.3
Preclinical Diagnosis

In 1983, discovery of genetic linkage of HD to *D4S10* and the subsequent evidence for absence of nonallelic heterogeneity in the disorder made presymptomatic or prenatal diagnostic HD testing a reality. The availability of a preclinical test poses the dilemma of diagnosing a deadly disease in a currently healthy individual without any option for effective treatment and pilot programs began cautiously (Meissen et al. 1988; Brandt et al. 1989; Wiggins et al. 1992) using recommended guidelines (World Federation of Neurology Research Group on Huntington's Disease 1990).

Measurement of the length of the *HD* CAG repeat constitutes a direct molecular test that is relatively inexpensive and can be offered to any individual at risk in a variety of situations, including prenatal testing, confirmation of a clinical diagnosis of HD and differential diagnosis. The new test has broader applicability and is less expensive but the uncertain intergenerational instability and reduced penetrance of alleles in the 34–to 39-unit range is problematic, as there is no a priori way of knowing whether these alleles will cause disease or whether gametic instability will increase the risk to unborn children. Discovery of the *HD* mutation has made genetic testing for individuals at risk more accessible, but the availability of an accurate predictive test is at best a mixed blessing without an effective treatment for the disorder.

5
HD Gene and Its Products

The *HD* gene is unique in the genome, spanning approximately 185 kb in 4p16.3, and is comprised of 67 exons (Fig. 3). It directs the synthesis of two major mRNA transcripts of 13.5 and 10.5 kb that arise from alternate polyadenylation sites in exon 67 (Ambrose et al. 1994). Both transcripts encode the same approximately 350 kDa protein and are found in low to moderate abundance in all fetal and adult cells and tissues that have been examined, including the neuronal targets of the *HD* mutation (Dure et al. 1995; Landwehrmeyer et al. 1995; Li et al. 1993; Strong et al. 1993; Stine et al. 1993).

The *HD* CAG repeat is located in exon 1, 17 codons from the ATG start of translation. Together with two penultimate invariant CAACAG codons, the polymorphic CAG segment encodes 8 to 36 glutamine residues. Immediately adjacent, a degenerate stretch of CAG and CCG codons, including the slightly polymorphic CCG repeat, encodes a broken array of approximately 40 residues that are predominantly proline. Further 3' in exon 58, the Δ2642 polymorphism that forms part of the "major" HD haplotype entails the absence of one of six glutamate codons on 5 % of normal and approximately 40 % of disease chromosomes (Ambrose et al. 1994).

5.1
Huntingtin

HD mRNA encodes an approximately 350 kD protein of unknown function called huntingtin that does not show similarity to previously reported sequences except in the low complexity amino terminal polyglutamine polyproline segment and a motif of unknown function ("HEAT" repeat) that is found in a number of unrelated proteins (Andrade and Bork 1995). The CAG repeat is translated into a polymorphic segment of glutamine residues located 17 amino acids from the extreme amino terminus of the protein (Jou et al. 1995; Ide et al. 1995; Persichetti et al. 1995; Trottier et al. 1995a).

Huntingtin is found in the cytoplasm of a wide variety of human, monkey, and rat neuronal and non-neuronal cells and tissues in a broad expression pattern that does not readily account for the neuronal specificity of the HD defect in humans (Gutekunst et al. 1995; Sharp et al. 1995; DiFiglia et al. 1995; Persichetti et al. 1995; Trottier et al. 1995a). Within the basal ganglia, however, the distribution of huntingtin immunoreactivity is heterogeneous, with the targets of the mutation, medium sized neurons, exhibiting variable levels of the protein, while large neurons not affected in the disease display low or undetectable levels of protein (Kosinski et al. 1997; Ferrante et al. 1997). These findings suggest that the pattern of neuronal cell death in the striatum is due to the relatively high levels of huntingtin expression found in selective medium sized neurons but does not explain why neuronal cell types in other regions of the brain that also express high levels of the protein are not affected in the disease. In neurons, huntingtin immunoreactivity is found in the cytoplasm throughout the cell body, in axons, dendrites, and perikarya, where it may be associated with microtubules (Gutekunst et al. 1995) and vesicles (DiFiglia et al. 1995), suggesting a role for the protein in trafficking or neurotransmission.

5.2
Evolutionary Conservation

HD gene homologues from mouse, rat and pufferfish encode proteins that are highly conserved across the entire approximately 3144 amino acid length of the protein (Barnes et al. 1994; Lin et al. 1994; Baxendale 1995; Schmitt et al. 1995; MacDonald et al. 1996). The least conserved region is the amino terminal polyglutamine proline rich stretch, although it is flanked by regions of near sequence identity (Fig. 6). In humans, this segment comprises 12–36 glutamines, while the rat homologue has eight, mouse seven, and the pufferfish gene four consecutive glutamine residues at this position. Except for the human gene, with its pure run of CAG, this segment of the protein is not polymorphic and is encoded by a broken array of CAG and CAA codons. The exact length and composition of this segment of the protein, therefore, is not likely to be crucial to huntingtin's normal activity.

```
Human   MATLEKLMKAFESLKSFQQQQQQQQQQQQQQQQQQQQQQQPPPPPPPPPPP...   48
Mouse   MATLEKLMKAFESLKSF............QQQQQQQPPPQPPPPPPPPPP      38
Rat     MATLEKLMKAFESLKSF..........QQQQQQQQQPPPQPPPPPPPPP.      38
Fugu    MATMEKLMKAFESLKSF..............QQQQGPP..........       24

Human   QLPQPPPQAQPLLPQPQPPPPPPPPPPPGPAVAEEPLHRPKKELSATKKDRVN  100
Mouse   QPPQPPPQGQ........PPPPPPPLPGPA..EEPLHRPKKELSATKKDRVN    80
Rat     QPPQPPPQGQ.........PPPPPPLPGPA..EEPLHRPKKELSATKKDRVN    79
Fugu    ...........................TAEEIVQRQKKEQATTKKDRVS      46
```

Fig. 6. The amino terminus of huntingtin and its homologues. The amino terminal 100 amino acids of huntingtin encoded by the human *HD* gene are depicted, with the mouse (Barnes et al. 1994; Lin et al. 1994), rat (Schmitt et al. 1995; MacDonald et al. 1996), and pufferfish (Fugu) (Baxendale et al. 1995) homologues depicted below to maximize sequence identity. The polymorphic glutamine proline rich segment is not well conserved but is flanked by regions of near sequence identity, suggesting that this portion of the protein is not important for huntingtin's activity

5.3
Function

Huntingtin's activity is not known, but its high sequence conservation in vertebrate evolution and ubiquitous, "housekeeping" expression pattern in all stages of development (Duyao et al. 1995; Zeitlin et al. 1995; Bhide et al. 1996; MacDonald et al. 1996) suggest that its cellular function is vital. Indeed, although it is not required for cell viability (Duyao et al. 1995; Zeitlin et al. 1995), the protein's activity is essential at the level of the whole organism. Elimination of huntingtin by homozygous *Hdh* inactivation produces embryos that fail during gastrulation at about embryonic day 7.5 before the elaboration of the nervous system (Duyao et al. 1995; Nasir et al. 1995; Zeitlin et al. 1995). The cause of this early characteristic embryonic lethality is not evident from *Hdh's* wide pattern of expression in the embryo, but understanding huntingtin's role in development may shed light on the protein's normal function.

Although the organism strictly requires huntingtin's function, the level of activity required to fulfill these needs may be relatively modest. Huntingtin activity can be reduced to 50 % of wild-type levels without deleterious consequences. In humans, heterozygous *HD* inactivation as a result of a chromosomal translocation is not associated with any abnormal phenotype (Ambrose et al. 1994), although the loss of as yet unidentified 4p16.3 genes in the vicinity produces 4p- and Wolf-Hirschhorn syndromes (Wright et al. 1997). Similarly, in the mouse, heterozygous *Hdh* inactivation does not lead to abnormal phenotypes (Duyao et al. 1995; Zeitlin et al. 1995).

6
Mechanism of Pathogenesis

Knowledge of the *HD* gene, its products, and the nature of the mutation does not provide immediate clues to the mechanism by which the expanded CAG produces the characteristic neuropathology and symptoms of HD. It could act at the level of the DNA, altering chromatin structure and transcription of 4p16.3 genes in the *HD* region. Similarly, the mRNA product of the disease allele might modulate normal nuclear RNA processing, cytoplasmic export, or mRNA translation. Alternatively, because the CAG repeat is translated into a lengthened polyglutamine segment, the defect could act at the level of huntingtin protein. The accumulated evidence does not formally exclude any of these scenarios, but favors the latter, focusing efforts on the discovery of a deleterious property that is peculiar to huntingtin's elongated amino terminal glutamine segment.

6.1
Mutant *HD* Gene and Its Products

Exploration of the consequences of the expanded CAG repeat on the *HD* gene and its products is necessarily constrained by the fact that HD is a dominant disorder that affects specific populations of neuronal cells typically in the adult years. However, it is apparent that the mutation elongates the CAG repeat tract at the 5' end of the *HD* mRNA, but does not dramatically alter the gene's ubiquitous pattern of expression. Transcripts from the normal and the disease allele are both expressed in the cells and tissues of heterozygous HD patients (Ambrose et al. 1994; Stine et al. 1995), arguing that the defect does not produce a loss of huntingtin function at the level of the HD gene or its mRNA products, and making it likely that the mutation acts via an altered protein product.

6.2
Mutant Huntingtin

The mutant *HD* allele encodes a structurally distinct huntingtin product that is found in HD cells and tissues in a pattern that mirrors that of its normal counterpart. The disease-causing versions of the protein can be distinguished because they possess a lengthened amino-terminal polyglutamine segment that confers altered migration on SDS-PAGE (Sharp et al. 1995; DiFiglia et al. 1995; Persichetti et al. 1995; 1996; Trottier et al. 1995a; Aronin et al. 1995) and enhanced reactivity to monoclonal reagents directed at lengthy glutamine arrays (Trottier et al. 1995b).

   In biochemical experiments, the alteration does not affect the protein's stability or cytoplasmic intracellular location (Persichetti et al. 1995; 1996), and there is no evidence of a selective accumulation of stable truncated huntingtin products in the cells and tissues of HD patients (Persichetti et al. 1996), although it is difficult to exclude subtle differences in postmortem tissue. Indeed, exploration of the mutant protein's pattern of expression and intracellular localization by immunocytochemistry in HD heterozygotes is complicated by the presence of huntingtin produced from the normal allele, and accurate studies await antibody reagents that specifically recognize the mutant protein in this format. There is, therefore, no direct evidence that the unstable expanded *HD* CAG repeat acts at the level of the mutant protein but similar observations in a growing number of dominantly inherited neurodegenerative disorders make this scenario even more compelling.

7
Other CAG Repeat Disorders

The same pathogenic process that operates in HD could ultimately lead to the specific patterns of neuronal cell loss which characterize other neurodegenerative disorders that are due to expanded, meiotically unstable CAG repeats encoding glutamine (Table 1). Indeed, the similarities in inheritance, dependence of disease onset and severity on CAG repeat length, and specificity of the mutation for neuronal cells implies that, in each of these disorders, the repeat triggers a common pathogenic mechanism (Gusella et al. 1997).

7.1
Spinal and Bulbar Muscular Atrophy

Kennedy's disease, or spinal bulbar muscular atrophy (SBMA), is caused by an expanded CAG repeat in the androgen receptor encoded at Xq11.2-q12 (LaSpada et al. 1991); in males, it produces a progressive loss of anterior horn cells in the spinal cord and concomitant muscle weakness (Fischbeck 1995; Brooks and Fischbeck 1995). Symptoms of affected males may, but do not always, include endocrine abnormalities suggestive of reduced receptor function (MacLean et al. 1995). Since inactivating mutations in the androgen receptor gene lead to testicular feminization, it is unlikely that the expanded CAG repeat causes SBMA by simple inactivation of the androgen receptor.

Table 1. Characteristics of glutamine encoding CAG repeat disorders

Disorder	Chromosome	Number of repeats		Gene	Protein product	Expression pattern	Brain region affected
		Normal	Affec-ted		Intracellular location		
Kennedy's disease (spinal bulbar muscular atrophy)	Xq11.2-q12	11–33	40–62	AR	Androgen receptor Cytoplasm, nucleus	Brain and periphery	Spinal cord, Anterior horn cells
Huntington's disease	4p16.3	11–34	37–121	HD	Huntingtin Cytoplasm	Brain and periphery	Caudate nucleus,cerebral cortex
Dentatorubral-pallidoluysian atrophy, Haw River syndrome	12p	8 35	49–79	DRPLA	Atrophin-1 Cytoplasm	Brain	Dentate nucleus,Globus pallidus
Spinocerebellar a-taxia 1	6p24	19–36	40–81	SCA1	Ataxin-1 Nucleus	Brain and periphery	Cerebellar cortex, inferior olive, cranial nerves
Spinocerebellar a-taxia 2	12q24.1	15–24	35–59	SCA2	Ataxin-2 Cytoplasm	Brain and periphery	Cerebellar cortex, inferior olive
Machado-Joseph disease, Spinocerebellar ataxia 3	14q32.1	12–40	67–82	MJD, SCA3	MJD1a protein Cytoplasm	Brain and periphery	Dentate, pontine, and vestibular nuclei, substantia nigra
Spinocerebellar a-taxia 6	19p13	4–16	21–27	CACNL 1A4	Isoform of α_{1A} subunit Brain voltage-gated calcium channel	Brain	Cerebellar cortex

7.2
Dentatorubral-Pallidolysian Atrophy

In dentatorubral-pallidoluysian atrophy (DRPLA), neuronal cell loss in both the dentatofugal and pallidofugal systems produces a distinctive set of clinical symptoms including both ataxia and choreoathetosis as well as myoclonus, epilepsy and dementia (Ikeuchi et al. 1995a; 1995b). *DRPLA*, on chromosome 12p, encodes a novel approximately 190 kDa cytoplasmic protein of unknown function, atrophin-1, which possesses a centrally located segment of polyglutamine (Koide et al. 1994; Nagafuchi et al. 1994; Yazawa et al. 1995; Margolis et al. 1996). Haw River syndrome, a disease in a single family in North Carolina with symptoms that closely parallel DRPLA except for the demyelination of the subcortical white matter tracts, is also caused by expansion of the CAG repeat at the *DRPLA* locus on 12p (Burke et al. 1994).

7.3
Spinocerebellar Ataxia

A number of clinically distinct forms of spinocerebellar ataxia (SCA) – types 1-3, 6, and probably 7 – are caused by CAG repeat expansion. In SCA1, an expanded CAG repeat in the novel 6p gene encoding ataxin-1 elongates a glutamine segment in the first third of a widely expressed, approximately 87 kDa nuclear protein of unknown function, producing progressive cerebellar ataxia, with muscle atrophy, decreased deep tendon reflexes, and loss of proprioception and vibration sense (Zhogbi and Orr 1995). In this case, the expanded repeat targets neurons in the cerebellum, both Purkinje cells and dentate nucleus neurons, the inferior olive, and cranial nerve nuclei III, IV, IX, X, and XII. Unlike HD, the larger *SCA1* normal alleles do not encode a pure CAG stretch, but rather one with CAT interruptions that are not found in the disease allele. Ataxin-1 protein expressed from normal alleles, therefore, has a glutamine segment punctuated by a histidine residue.

A clinically similar disorder, SCA2, is characterized by progressive ataxia with variable peripheral involvement, including lid retraction, dementia, axonal sensory neuropathy and degeneration of the posterior columns associated with neuronal cell loss in the cerebellum, inferior olives, pontine nuclei and substantia nigra. The disease is caused by expansion of a CAG repeat in a novel gene at 12q24.1 that extends a polyglutamine segment in a novel cytoplasmic approximately 150 kDa protein (Pulst et al. 1996; Sanpei et al. 1996; Imbert et al. 1996).

In SCA3 and Machado-Joseph disease (MJD), allelic disorders due to CAG repeat expansion in the same novel cytoplasmic, approximately 47kDa MJD1a protein encoded by a gene at 14q32.1, the defect targets neurons in the spinocerebellar tracts, sparing neurons in the inferior olive and pontine nuclei that are affected in SCA1 and SCA2 (Kawaguchi et al. 1994; Schöls et al. 1995a; 1995b; Ikeda et al. 1996).

SCA6, with progressive cerebellar and brain stem dysfunction, is due to a CAG repeat in an isoform of the α_{1A} subunit of a brain voltage-gated calcium channel (*CACNL1A4*) on 19p13 (Zhuchenko et al. 1997). Interestingly, in this case the expanded repeat is stably inherited and encodes a polyglutamine segment of modest length that falls squarely in the range found on normal chromosomes in the other diseases. SCA7 is a genetically and clinically distinct cerebellar ataxia with retinal degeneration that is associated with an expanded triplet repeat and polyglutamine segment in a 130 kDa nuclear protein that has not yet been isolated (Trottier et al. 1995b; Lindblad et al. 1996).

7.4
A Common Pathogenic Mechanism?

In several of these CAG repeat disorders, there is a correlation between the size of the expanded allele and age at onset of disease comparable to that found in HD (Fig. 7; Gusella et al. 1997). Indeed, the SCA1 and HD data are indistinguishable when age at onset is plotted against CAG repeat length. By contrast, DRPLA and MJD/SCA3 data both display slopes that are slightly steeper and shifted toward higher repeat lengths, whereas the SCA2 data has a similar shape but lies to the left of the HD curve. Thus the degree of CAG expansion required to cause a detectable neurologic deficit at any given age differs in each of these diseases, and in DRPLA, MJD/SCA3, and SCA2 the addition of each CAG repeat has a greater impact on reducing age at onset of clinical symptoms that in HD or SCA1. It seems reasonable, therefore, that the expanded CAG repeat in each case triggers a common mechanism, but that the context in which the defect is presented to the cell determines the neuronal target and sensitivity to CAG length. If there is a gradual decline in neuronal function until a threshold for onset of clinical symptoms is crossed, then the dissimilar shapes of the onset curves in each disorder could reflect differences in the extent of neuronal loss required in distinct neuronal populations before symptoms are elicited as well as the context dependent differences in the rate of functional decline elicited by the CAG repeat. This idea is supported by the finding that the juvenile onset form of each disease is associated with neuronal cell loss in regions of the brain that are spared in the more typical adult onset disorder. Thus, within the same context the CAG expansion is capable of causing the death of different neuronal cells, but each population declines at a different rate determined by the length of the repeat.

Another conspicuous difference is the age at onset exhibited by individuals that are homozygous for the mutant allele in HD and the other disorders (Fig. 7). A second copy of the disease allele does not significantly alter the disease process in HD (Myers et al. 1989), SCA1 (Goldfarb et al. 1996) or SCA2 (Sanpei et al. 1996) but produces a more severe disease with an earlier age at onset in both DRPLA (Sato et al. 1995) and SCA3/MJD (Lang et al. 1994; Takiyama et al. 1995; Lerer et al. 1996). In HD, SCA1, and SCA2, therefore, the presence of a

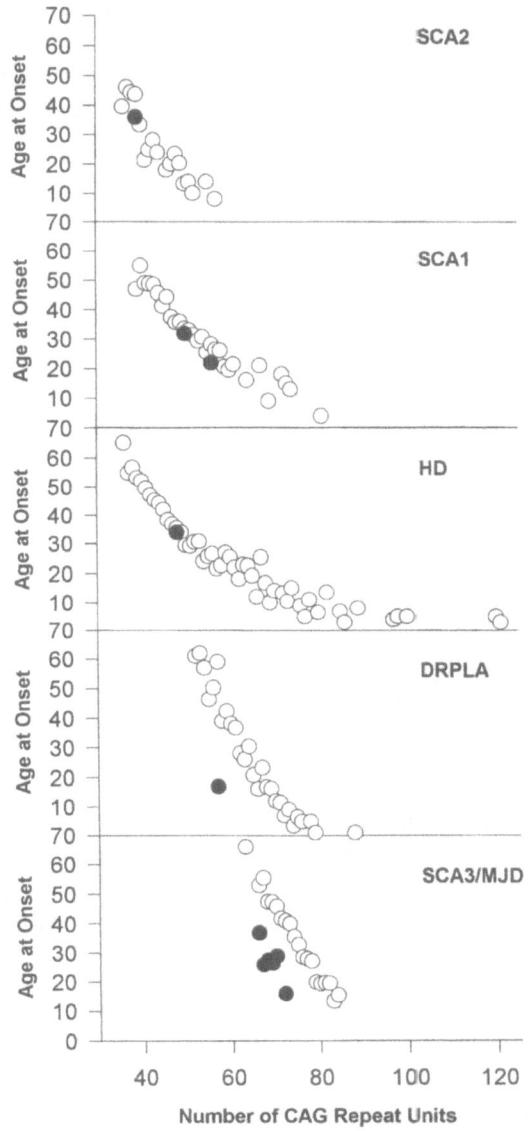

Fig. 7. Comparison of the relationship of CAG repeat length and age at onset in HD and other CAG repeat diseases. The mean age at onset of neurologic symptoms is plotted against length of the expanded CAG repeat on the disease chromosome in heterozygous (*open circles*) and homozygous (*filled circles*) spinocerebellar ataxia (SCA) type 2, SCA1, HD, dentatorubral-pallidoluysian atrophy (DRPLA), and SCA3/Machado-Joseph disease (MJD) patients (Gusella et al. 1997). The age at onset of disease symptoms in each case is highly correlated with CAG repeat length. HD and SCA1 have very similar curves, but those for SCA2 and DRPLA and SCA3/MJD are shifted to the left and right, respectively, demonstrating that the context of the repeat impacts the rate at which the defect causes symptoms. Like HD, the second disease allele in homozygous individuals does not produce altered clinical onset in SCA1 and SCA2, but two doses of the expanded repeat combine to hasten onset in DRPLA and SCA3/MJD, suggesting that the disease process is not saturated in heterozygotes

single mutant gene is sufficient to saturate the pathogenic process. By contrast, in DRPLA and MJD/SCA3, the pathogenic mechanism is not saturated in typical heterozygotes, having the capacity to be accelerated by the addition of a second CAG expansion allele.

8
Prospects

The discovery of the *HD* gene's unstable, expanded CAG mutation and characterization of its behavior has provided the means to accurately diagnose the disorder, thereby improving clinical management of the disease. However, there are no interventions for the disorder, and discovery of the defect holds the unfulfilled promise that elucidation of the pathogenic mechanism will lead to the development of therapies.

It is probable that the *HD* mutation and the expanded CAG repeats in the other disorders trigger a similar pathogenic processes. Delineation of the mechanism in one of these diseases, therefore, will ultimately lead to an understanding of the others, revealing the basis of the astounding neuronal specificity of the mutation in each disorder. The mutant protein products are likely to trigger the disease process in a manner that is independent of their unrelated inherent activities, although loss of these normal functions may contribute to aspects of the disease phenotype. Differences in the capacity of the mutant proteins to trigger the pathogenic mechanism may depend on concentration, localization, normal function, and constraints imposed by the protein's structure and the relative location of the polyglutamine segment within it. The disease process may involve pathways that are common to many cell types or, alternatively, may be the result of physiological processes peculiar to neuronal cells. It may involve the formation of a toxic product from the mutant protein, perhaps by transglutamination (Green 1993) or cleavage by enzymes in the apoptotic pathway (Goldberg et al. 1996), or the mutant protein may alter some aspect of energy metabolism triggering excitotoxic pathways (Beal 1994).

The ability to compare a number of disorders caused by unstable CAG repeats affords one of the most promising routes to accelerating the task of discovering the detailed steps that lead from CAG expansion to specific neuronal cell death in HD. This knowledge, it is hoped, will hasten the development of rational therapeutic interventions for all of these devastating, untreatable CAG expansion disorders.

Acknowledgments. The author is grateful to colleagues Jacqueline White, Peter Faber, Teresa Calzonetti, Vanessa Wheeler, Jayalakshmi Srinidhi, Ingrid Pribill, Jianmin Chen, Vladimir Vrbanac, Caterina Dompe and Francesca Persichetti for their dedicated efforts on behalf of HD research. Special thanks to James Gusella for continued lively discussion of the ideas advanced in this manuscript. The author's research is supported by grants from the National Institutes of Health (NS16367, NS32765), Bristol-Myers Squibb Inc., the Hereditary Disease Foundation, The Huntington's Disease Society of America and the Foundation for the Care and Cure of Huntington's disease.

References

Albin RL, Reiner A, Anderson KD, Dure LS 4th, Handelin B, Balfour R, Whetsell WO Jr, Penney JB, Young AB (1992) Preferential loss of striato-external pallidal projection neurons in presymptomatic Huntington's disease. Ann Neurol 31:425–430

Ambrose CM, Duyao MP, Barnes G, Bates GP, Lin CS, Srinidhi J, Baxendale S, Hummerich H, Lehrach H, Altherr M, Wasmuth JJ, Buckler A, Church D, Housman D, Berks M, Micklem G, Durbin R, Dodge A, Read A, Gusella JF, MacDonald ME (1994) Structure and expression of the Huntington's disease gene: evidence against simple inactivation due to an expanded CAG repeat. Somat Cell Mol Genet 20:27–38

Andrade MA, Bork P (1995) HEAT repeats in the Huntington's disease protein Nat Genet 11:115–116

Andrew SE, Goldberg YP, Kremer B, Telenius H, Theilmann J, Adam S, Starr E, Squitieri F, Lin B, Kalchman MA, Graham RK, Hayden MR (1993) The relationship between trinucleotide (CAG) repeat length and clinical features of Huntington's disease. Nat Genet 4:398–403

Andrew SE, Goldberg YP, Theilmann J, Zeisler J, Hayden MR (1994a) A CCG repeat polymorphism adjacent to the CAG repeat in the Huntington disease gene: implications for diagnostic accuracy and predictive testing. Hum Mol Genet 3:65–67

Andrew SE, Goldberg YP, Kremer B, Squitieri F, Theilmann J, Zeisler J, Telenius H, Adam S, Almquist E, Anvret M, Lucotte G, Stoessl AJ, Campenella G, Hayden MR (1994b) Huntington disease without CAG expansion: phenocopies or errors in assignment? Am J Hum Genet 54:852–863

Aronin N, Chase K, Young C, Sapp E, Schwartz C, Matta N, Kornreich R, Landwehrmeyer B, Bird E, Beal MF, Vonsattel J-P, Smith T, Carraway R, Boyce FM, Young AB, Penney JB, DiFiglia M (1995) CAG expansion affects the expression of mutant huntingtin in the Huntington's disease brain. Neuron 15:1193–1201

Barnes GT, Duyao MP, Ambrose CM, McNeil S, Persichetti F, Srinidhi J, Gusella JF, MacDonald ME. Mouse Huntington's disease gene homolog (Hdh) (1994)Somat Cell Mol Genet 20:87–97

Barron LH, Warner JP, Porteous M, Holloway S, Simpson S, Davidson R, Brock DJH (1993) A study of the Huntington's disease associated trinucleotide repeat in the Scottish population. J Med Genet 30:1003–1007

Baxendale S, Abdulla S, Elgar G, Buck D, Berks M, Micklem G, Durbin R, Bates GP, Brenner S, Beck S, Lehrach H (1995) Comparative sequence analysis of the human and pufferfish Huntington's disease genes. Nat Genet 10:67–76

Beal MF (1994) Huntington's disease, energy and excitotoxicity Neurobiol Aging 15:275–276

Bhide PF, Day M, Sapp E, Schwarz C, Sheth A, Kim J, Young AB, Penney J, Golden J, Aronin N, DiFiglia M (1996) Expression of normal and mutant huntingtin in the developing brain. J Neurosci 16:5523–5535

Biancalana V, Serville F, Pommier J, Julien J, Hanauer A, Mandel JL (1992) Moderate instability of the trinucleotide repeat in spino-bulbar muscular atrophy. Hum Mol Genet 1:255–258

Bird ED, Caro AJ, Pilling JB (1974) A sex related factor in the inheritance of Huntington's chorea. Ann Hum Genet 37:255–260

Bozza A, Malagu S, Calzolari E, Novelletto A, Pavoni M, del Senno L (1995) Expansion of a (CAG)n repeat region in a sporadic case of HD. Acta Neurol Scand 92:132–134.

Brandt J, Quaid KA, Folstein SE, Garber P, Maestri NE, Abbott MH, Slavney PR, Franz ML, Kasch L, Kazazian HH (1989) Presymptomatic diagnosis of delayed-onset disease with linked DNA markers: The experience in Huntington's disease. J Am Med Assoc 261:3108–3114

Brandt J, Bylsma FW, Gross R, Stine OC, Ranen N, Ross CA (1996) Trinucleotide repeat length and clinical progression in Huntington's disease. Neurology 46:527–531

Brooks BP and Fischbeck KH (1995) Spinal and bulbar muscular atrophy: a trinucleotide-repeat expansion neurodegenerative disease. Trends Neurosci 18:459–461

Burke JR, Wingfield MS, Lewis KE, Roses AD, Lee JE, Hulette C, Pericak-Vance MA, Vance JM (1994) The Haw River Syndrome: dentatorubropallidoluysian atrophy (DRPLA) in an African-American family. Nat Genet 7:521–524

Cancel G, Abbas N, Stevanin G, Durr A, Chneiweiss H, Neri C, Duyckaerts C, Penet C, Cann HM, Agid Y, Brice A (1995) Marked phenotypic heterogeneity associated with expansion of a CAG repeat sequence at the spinocerebellar ataxia 3/Machado-Joseph disease locus. Am J Hum Genet 57:809–816

Claes S, Van Zand K, Legius E, Dom R, Malfroid M, Baro F, Godderis J, Cassiman JJ (1995) Correlations between triplet repeat expansion and clinical features in Huntington's disease. Arch Neurol 52:749–753

Conneally PM, Haines JL, Tanzi RE, Wexler NS, Penchaszadeh GK, Harper PS, Folstein SE, Cassiman JJ, Myers RH, Young AB, Hayden MR, Falek A, Tolosa ES, Crespi S, Di Maio L, Holmgren G, Anvret M, Kanazawa I, Gusella JF (1989) Huntington disease: No evidence for locus heterogeneity. Genomics 5:304–308

Craufurd D, Dodge (1993) AMutation size and age at onset in Huntington's disease. J Med Genet 30:1008–1011

Davis MB, Bateman D, Quinn NP, Marsden CD, Harding AE (1994) Mutation analysis in patients with possible but apparently sporadic Huntington's disease. Lancet. 344:714–717

de la Monte SM, Vonsattel JP, Richardson EP, Jr (1988) Morphometric demonstration of atrophic changes in the cerebral cortex, white matter, and neostriatum in Huntington's disease. J Neuropathol Exp Neurol 47:516–525

De Rooij KE, De Koning, Gans PA, Skraastad MI, Belfroid RD, Vegter-Van Der Vlis M, Roos RA, Bakker E, Van Ommen GJ, Den Dunnen JT, Losekoot M (1993) Dynamic mutation in Dutch Huntington's disease patients: increased paternal repeat instability extending to within the normal size range. J Med Genet 30:996–1002

De Rooij KE, De Konig Gans PA, Roos RA, Van Ommen GJ, Den Dunnen JT (1995) Somatic expansion of the (CAG)n repeat in Huntington disease brains. Hum Genet 95:270–274

DiFiglia M, Sapp E, Chase K, Schwarz C, Meloni, Young C, Martin E, Vonsattel JP, Carraway R, Reeves SA, Boyce FM, Aronin N (1995) Huntingtin is a cytoplasmic protein associated with vesicles in human and rat brain neurons. Neuron 14:1075–1081

Dode C, Durr A, Pecheux C, Mouret JF, Belal S, Bachner L, Agid Y, Kaplan JC, Brice A, Feingold J (1993) Huntington's disease in French families: CAG repeat expansion and linkage disequilibrium analysis. C R Acad Sci III 316:1374–1380

Dubourg O, Durr A, Cancel G, Stevanin G, Chneiweiss H, Penet C, Agid Y, Brice A (1995) Analysis of the SCA1 CAG repeat in a large number of families with dominant ataxia: clinical and molecular correlations. Ann Neurol 37:176–180

Dure LS, Landwehrmeyer GB, Golden J, McNeil SM, Ge P, Aizawa H, Huang Q, Ambrose CM, Duyao MP, Bird ED, DiFiglia M, Gusella JF, MacDonald ME, Penney JB, Young AB, Vonsattel J-P (1994) IT15 expression in fetal human brain. Brain Res 659: 33–41

Durr A, Dode C, Hahn V, Pecheux C, Pillon B, Feingold J, Kaplan JC, Agid Y, Brice A (1995) Diagnosis of "sporadic" Huntington's disease. J Neurol Sci 129:51–55

Durr A, Stevanin G, Cancel G, Duyckaerts C, Abbas N, Didierjean O, Chneiweiss H, Benomar A, Lyon-Caen O, Julien J, Serdaru M, Penet C, Agid Y, Brice A (1996) Spinocerebellar ataxia 3 and Machado-Joseph disease: clinical, molecular, and neuropathological features. Ann Neurol 39:490–499

Duyao M, Ambrose C, Myers R, Novelletto A, Persichetti F, Frontali M, Folstein S, Ross C, Franz M, Abbott M, Gray J, Conneally PM, Young A, Penney J, Hollingsworth Z, Shoulson I, Lazzarini AM, Falek A, Koroshetz W, Sax DS, Bird E, Vonsattel JP, Bonilla E, Alvir J, Bickham Conde J, Cha JH, Dure L, Gomez F, Ramos M, Sanchez-Ramos J, Snodgrass SR, de Young M, Wexler NS, Moscowitz C, Penchaszadeh G, MacFarlane H, Anderson MA, Jenkins B, Srinidhi J, Barnes G, Gusella JF, MacDonald ME (1993) Trinucleotide repeat length instability and age of onset in Huntington's disease. Nat Genet 4:387–392

Duyao MP, Auerbach A, Ryan A, Persichetti F, Barnes GT, McNeil SM, Ge P, Vonsattel J-P, Gusella JF, Joyner AL, MacDonald ME (1995) Inactivation of the mouse Huntington's disease gene homolog Hdh. Science 269: 407–410

Ferrante RJ, Kowall NW, Beal MF, Richardson EP Jr, Bird ED, Martin JB (1985) Selective sparing of a class of striatal neurons in Huntington's disease. Science 230:561–563

Ferrante RJ, Gutekunst CA, Persichetti F, Kowall N, Gusella JF, Beal MF, MacDonald ME, Hersch SM (1997) Heterogeneous topographic and cellular distribution of huntingtin expression in the normal human neostriatum. J Neurosci 17:3052-3063

Fischbeck KH (1995) The expanded trinucleotide repeat in Kennedy's disease. Proc Assoc Am Phys 107:228-230

Folstein S (1989) Huntington's disease a disorder of families. The Johns Hopkins Press, Baltimore, pp13-64.

Furtado S, Suchowersky O, Rewcastle B, Rewcastle B, Graham L, Klimek ML, Garber A (1996) Relationship between trinucleotide repeats and neuropathological changes in Huntington's disease. Ann Neurol 39:132-136

Gilliam TC, Tanzi RE, Haines JL, Bonner TI, Faryniarz AG, Hobbs WJ, MacDonald ME, Cheng SV, Folstein SE, Conneally PM, Wexler NS, Gusella JF (1987) Localization of the Huntington's disease gene to a small segment of chromosome 4 flanked by D4S10 and the telomere. Cell 50:565-571

Goldberg YP, Kremer B, Andrew SE, Theilmann J, Graham RK, Squitieri F, Telenius H, Adam S, Sajoo A, Starr E, Heiberg A, Wolff G, Hayden MR (1993) Molecular analysis of new mutations for Huntington's disease: intermediate alleles and sex of origin effects. Nat Genet 5:174-179

Goldberg YP, Nicholson DW, Rasper DM, Kalchman MA, Koide HB, Graham RK, Bromm M, Kazemi-Esfarjani P, Thornberry NA, Vaillancourt JP, Hayden MR (1996) Cleavage of huntingtin by apopain, a proapoptotic cysteineprotease, is modulated by the polyglutamine tract. Nat Genet 13: 442-449

Goldfarb LG, Vasconcelos O, Platonov FA, Lunkes A, Kipnis V, Kononova S, Chabrashvili T, Vladimirtsev VA, Alexeev VP, Gajdusek DC (1996) Unstable triplet repeat and phenotypic variability of spinocerebellar ataxia type 1. Ann Neurol 39:500-506.

Graveland GA, Williams RS, DiFiglia M (1985) Evidence for degenerative and regenerative changes in neostriatal spiny neurons in Huntington's disease. Science 227:770-773

Green H (1993) Human genetic diseases due to codon reiteration: Relationship to an evolutionary mechanism. Cell 74:955-956

Gusella JF, MacDonald ME (1995) Huntington's disease. Semin Cell Biol 1995; 6:21-28

Gusella JF, MacDonald ME (1996) Trinucleotide instability: a repeating theme in human inherited disorders. Annu Rev Med 47:201-209

Gusella JF, Wexler NS, Conneally PM, Naylor SL, Anderson MA, Tanzi RE, Watkins PC, Ottina K, Wallace MR, Sakaguchi AY, Young AB, Shoulson I, Bonilla E, Martin JB (1983) A polymorphic DNA marker genetically linked to Huntington's disease. Nature 306:234-238

Gusella JF, Tanzi RE, Anderson MA, Hobbs W, Gibbons K, Raschtchian R, Gilliam TC, Wallace MR, Wexler NS, Conneally PM (1984) DNA markers for nervous system diseases. Science 225:1320-1326

Gusella JF, Tanzi RE, Bader PI, Phelan MC, Stevenson R, Hayden MR, Hofman KJ, Faryniarz AG, Gibbons K (1985) Deletion of Huntington's disease-linked G8 (D4S10) locus in Wolf-Hirschhorn syndrome. Nature 318: 75-78

Gusella JF, Persichetti F, MacDonald ME (1997) The genetic defect causing Huntington's disease: repeated in other contexts? Mol Med 4:238-246.

Gutekunst C-A, Levey AI, Heilman CJ, Whaley WL, Hong Y, Nash NR, Rees HD, Madden JJ, Hersch SM (1985) Identification and localization of huntingtin in brain and human lymphoblastoid cell lines with anti-fusion protein antibodies. Proc Natl Acad Sci (USA) 92:8710-8714

Haberhausen G, Damian MS, Leweke F, Muller U (1995) Spinocerebellar ataxia, type 3 (SCA3) is genetically identical to Machado-Joseph disease (MJD). J Neurol Sci 132:71-75

Harper PS (1992)The epidemiology of Huntington's disease. Hum Genet 89:365-376

Hedreen JC, Peyser CE, Folstein SE, Folstein SE, Ross CA (1991) Neuronal loss in layers V and VI of cerebral cortex in Huntington's disease. Neurosci Lett 133:257-261

Higgins JJ, Nee LE, Vasconcelos O, Ide SE, Lavedan C, Goldfarb LG, Polymeropoulos MH (1996) Mutations in American families with spinocerebellar ataxia (SCA) type 3: SCA3 is allelic to Machado-Joseph disease. Neurology 46:208-213

Huntington G On chorea (1872) Med Surg Rep 26:317-321

Huntington's Disease Collaborative Research Group (1993) A novel gene containing a trinucleotide repeat that is expanded and unstable on Huntington's disease chromosomes. Cell 72:971–983

Ide K, Nukina N, Masuda N, Goto J, Kanazawa I (1995) Abnormal gene product identified in Huntington's disease lymphocytes and brain. Biochem Biophys Res Commun 209: 1119–1125

Ikeda H, Yamaguchi M, Sugai S, Aze Y, Narumiya S, Kakizuka A (1996) Expanded polyglutamine in the Machado-Joseph disease protein indices cell death in vitro and in vivo. Nat Genet 13: 196–202

Ikeuchi T, Onodera O, Oyake M, Koide R, Tanaka H, Tsuji S (1995a) Dentatorubral-pallidoluysian atrophy (DRPLA): close correlation of CAG repeat expansions with the wide spectrum of clinical presentations and prominent anticipation. Sem Cell Biol 6:37–44

Ikeuchi T, Koide R, Onodera O, Tanaka H, Oyake M, Takano H, Tsuji S (1995b) Dentatorubral-pallidoluysian atrophy (DRPLA). Molecular basis for wide clinical features of DRPLA. Clin Neurosci 3:23–27

Illarioshkin SN, Igarashi S, Onodera O, Markova ED, Nikolskaya NN, Tanaka H, Chabrashwili TZ, Insarova NG, Endo K, Ivanova-Smolenskaya IA, Tsuji S (1994) Trinucleotide repeat length and rate of progression of Huntington's disease. Ann Neurol 36:630–635

Imbert G, Saudou F, Yvert G, Devys D, Trottier Y, Garnier JM, Weber C, Mandel JL, Cancel G, Abbas N, Durr A, Didierjean O, Stevanin G, Agid Y, Brice A (1996) Cloning of the gene for spinocerebellar ataxia 2 reveals a locus with high sensitivity to expanded CAG/glutamine repeats. Nat Genet 14:285–291

International Huntington Association and World Federation of Neurology Research Group on Huntington's Chorea (1994) Guidelines for the molecular genetics predictive test in Huntington's disease. Neurology 44:1533–1536

Jou YS, Myers RM (1995). Evidence from antibody studies that the CAG repeat in the Huntington disease gene is expressed in the protein. Hum Mol Genet 4, 465–469

Kawaguchi Y, Okamoto T, Taniwaki M, Aizawa M, Inoue M, Katayama S, Kawakami H, Nakamura S, Nishimura M, Akiguchi I, Kimura J, Narumiya S, Kakizuka A (1994) CAG expansions in a novel gene for Machado-Joseph disease at chromosome 14q32.1. Nat Genet 8:221–228

Kieburtz K, MacDonald M, Shih C, Feigin A, Steinberg K, Bordwell K, Zimmerman C, Srindihi J, Sotack J, Gusella J, Shoulson I (1994) Trinucleotide repeat length and progression of illness in Huntington's disease. J Med Genet 31:872–874

Koide R, Ikeuchi T, Onodera O, Tanaka H, Igarashi S, Endo K, Takahashi H, Kondo R, Ishikawa A, Hayashi T, Saito M, Tomoda A, Miike T, Naito H, Ikuta F, Tsuji S (1994) Unstable expansion of CAG repeat in hereditary dentatorubral-pallidoluysian atrophy (DRPLA). Nat Genet 6:9–13

Komure O, Sano A, Nishino N, Yamauchi N, Ueno S, Kondoh K, Sano N, Takahashi M, Murayama N, Kondo I, Nagafuchi S, Yamada M, Kanazawa I (1995) DNA analysis in hereditary dentatorubral-pallidoluysian atrophy: correlation between CAG repeat length and phenotypic variation and the molecular basis of anticipation. Neurology 45:143–149

Koskinski C, Cha JH, Young AB, MacDonald ME, Persichetti F, Dawson TM, Dawson V, Penney JB, Standaert D (1997) Huntingtin in the neostriatum: selective accumulation in vulnerable neurons. Ann Neurol 144:239–247

Kremer B, Squitieri F, Telenius H, Andrew SE, Theilmann J, Spence N, Goldberg YP, Hayden MR (1993) Molecular analysis of late onset Huntington's disease. J Med Genet 30:991–995

Kremer B, Goldberg P, Andrew SE, Theilmann J, Telenius H, Zeisler J, Squitieri F, Lin B, Bassett A, Almqvist E, Bird T, Hayden MR (1994) A worldwide study of the Huntington's disease mutation. The sensitivity and specificity of measuring CAG repeats. N Engl J Med 330:1401–1406

Landegent JE, Jansen IN, De Wal N, Fisser-Groen YM, Bakker E, Van Der Ploeg M, Pearson PL (1986) Fine mapping of the Huntington disease linked D4S20 locus by non-radioactive in situ hybridization. Hum Genet 3:354–357

Landwehrmeyer GB, McNeil SM, Dure LS, 4th, Ge P, Aizawa H, Huang Q, Ambrose CM, Duyao MP, Bird ED, Bonilla E, de Young M, Avila-Gonzales AJ, Wexler NS, DiFiglia M, Gusella JF, MacDonald ME, Penney JB, Young AB, Vonsattel J-P (1995) Huntington's disease gene: regional and cellular expression in brain of normal and affected individuals. Ann Neurol 37:218–230

Lang AE, Rogaeva EA, Tsuda T, Hutterer J, St George-Hyslop P (1994) Homozygous inheritance of the Machado-Joseph disease gene. Ann Neurol 36:443–447

LaSpada AR, Wilson EM, Lubahn DB, Harding AE, Fishbeck H (1991) Androgen receptor gene mutations in X-linked spinal and bulbar muscular atrophy. Nature 352:77–79

LaSpada AR, Roling DB, Harding AE, Warner CL, Speigel R, Hausmanowa-Petrusewicz I, Yee W-C, Fischbeck KH (1992) Meiotic stability and genotype-phenotype correlation of the trinucleotide repeat in X-linked spinal and bulbar muscular atrophy. Nat Genet 2:301–304

Leeflang E, Zhang L, Tavare S, Hubert R, Srinidhi J, MacDonald ME, Myers R.H., de Young Margot, Wexler NS, Gusella JF, Arnheim N (1995) Single sperm analysis of the trinucleotide repeats in the Huntington's disease gene: quantification of the mutation frequency spectrum. Hum Molec Genet 4: 1519–1526

Lerer I, Merims D, Abeliovich D, Zlotogora J, Gadoth N (1996) Machado-Joseph disease: correlation between the clinical features, the CAG repeat length and homozygosity for the mutation. Eur J Hum Genet 4:3–7

Li SH, Schilling G, Young WS, 3D, Li XJ, Margolis RL, Stine OC, Wagster MV, Abbott MH, Franz ML, Ranen NG, Folstein SE, Hedreen JC, Ross CA (1993) Huntington's disease gene (IT15) is widely expressed in human and rat tissues. Neuron 11:985–993

Lin B, Nasir J, MacDonald H, Hutchinson G, Graham RK, Rommens JM, Hayden MR 1994) Sequence of the murine Huntington disease gene: evidence for conservation, alternate splicing and polymorphism in a triplet (CCG) repeat [published erratum appears in Hum Mol Genet 1994;3:530]. Hum Mol Genet 3:85–92

Lindblad K, Savontaus ML, Stevanin G, Holmberg M, Digre K, Zander C, Ehrsson H, David G, Benomar A, Nikoskelainen E, Trottier Y, Holmgren G, Ptacek LJ, Anttinen A, Brice A, Schalling M 1996) An expanded CAG repeat sequence in spinocerebellar ataxia type 7. Genome Res 6:965–971

Lucotte G, Aouizerate A, Loreille O, Gerard N, Turpin JC (1994) Trinucleotide repeat elongation in the huntingtin gene in Huntington's disease patients from 85 French families. Genet Couns 5:321–328

MacDonald ME, Novelletto A, Lin C, Tagle D, Barnes G, Bates G, Taylor S, Allitto B, Altherr M, Myers R, Lehrach H, Collins FS, Wasmuth JJ, Frontali M, Gusella JF (1992) The Huntington's disease candidate region exhibits many different haplotypes. Nat Genet 1:99–103

MacDonald ME, Barnes G, Srinidhi J, Duyao MP, Ambrose CM, Myers RH, Gray J, Conneally PM, Young A, Penney J, Shoulson I, Hollingsworth Z, Koroshetz W, Bird E, Vonsattel JP, Bonilla E, Moskowitz C, Penchaszadeh G, Brzustowicz L, Alvir J, Bickham Conde J, Cha J-H, Dure L, Gomez F, Ramos-Arroyo M, Sanchez-Ramos J, Snodgrass SR, de Young M, Wexler NS, Mac-Farlane H, Anderson MA, Jenkins B, Gusella JF (1993a) Gametic but not somatic instability of CAG repeat length in Huntington's disease. J Med Genet 30:982–986

MacDonald ME, Ambrose CM, Duyao MP, Gusella JF (1993b) Capturing a CAGey Killer. In: Davis KE, Warren S (eds) Genome Analysis vol 7: Genome rearrangement and instability. Cold Spring Harbor Press New York pp25–41

MacDonald ME, Duyao M, Calzonetti T, Auerbach A, Ryan A, Barnes G, White JK, Auerbach W, Vonsattel J-P, Gusella JF, Joyner AL (1996) Targeted inactivation of the mouse Huntington disease homologue Hdh. Cold Spring Harbor Symp Quant Bio 61:627–638

Maciel P, Gaspar C, DeStefano AL, Silveira I, Coutinho P, Radvany J, Dawson DM, Sudarsky L, Guimaraes J, Loureiro JE, Nezarati MM, Corwin LI, Lopes-Cencdes I, Rooke K, Rosenberg R, MacLeod P, Farrer LA, Sequeiros J, Rouleau GA (1995) Correlation between CAG repeat length and clinical features in Machado-Joseph disease. Am J Hum Genet 57:54–61

MacLean HE, Choi WT, Rekaris G, Warne GL, Zajac JD (1995) Abnormal androgen receptor binding affinity in subjects with Kennedy's disease (spinal and bulbar muscular atrophy). J Clin Endocrinol Metab 80:508–516

Margolis RL, Li SH, Young WS, Wagster MV, Stine OC, Kidwai AS, Ashworth RG, Ross CA (1996) DRPLA gene (atrophin-1) sequence and mRNA expression in human brain. Brain Res Mol Brain Res 36: 219–226

Martin JB, Gusella JF (1986) Huntington's disease: Pathogenesis and management. N Eng J Med 315:1267–1276

Maruyama H, Nakamura S, Matsuyama Z, Sakai T, Doyu M, Sobue G, Seto M, Tsujihata M, Oh-i T, Nishio T, Sunohara N, Takahashi R, Hayashi M, Nishino I, Ohtake T, Oda T, Nishimura M, Saida T, Matsumoto H, Baba M, Kawaguchi Y, Kakizuka A, Kawakami H (1995) Molecular features of the CAG repeats and clinical manifestation of Machado-Joseph disease. Hum Mol Genet 4:807–812

Masuda N, Goto J, Murayama N, Watanabe M, Kondo I, Kanazawa I (1995) Analysis of triplet repeats in the huntingtin gene in Japanese families affected with Huntington's disease. J Med Genet 32:701–705

Matilla T, Volpini V, Genis D, Rosell J, Corral J, Davalos A, Molins A, Estivill X (1993) Presymptomatic analysis of spinocerebellar ataxia type 1 (SCA1) via the expansion of the SCA1 CAG-repeat in a large pedigree displaying anticipation and parental male bias. Hum Mol Genet 2:2123–2128

Matilla T, McCall A, Subramony SH, Zoghbi HY (1995) Molecular and clinical correlations in spinocerebellar ataxia type 3 and Machado-Joseph disease. Ann Neurol 38:68–72

McNeil SM, Novelletto A, Srinidhi J, Barnes G, Kornbluth I, Altherr MR, Wasmuth JJ, Gusella JF, MacDonald ME, Myers RH (1997) Reduced penetrance of the Huntington̄s disease mutation. Hum Mol Genet 6:775–779

Meissen GJ, Myers RH, Mastromauro CA, Koroshetz WJ, Klinger KW, Farrer LA, Watkins PA, Gusella JF, Bird ED, Martin JB (1988) Predictive testing for Huntington's disease with use of a linked DNA marker. N Engl J Med 318:535–542

Merritt AD, Conneally PM, Rahman NF, Drew AL (1969) Juvenile Huntington's chorea. In Barbeau A, Brunette TR (eds) Progress in Neurogenetics. Excerpta Medica Foundation, Amsterdam pp645– 650

Myers RH, Leavitt J, Farrer LA, Jagadeesh J, McFarlane H, Mark RJ, Gusella JF (1989) Homozygote for Huntington's disease. Am J Hum Genet 45:615–618

Myers RH, MacDonald ME, Koroshetz WJ, Duyao MP, Ambrose CM, Taylor SAM, Barnes G, Srinidhi J, Lin CS, Whaley WL, Lazzarini AM, Schwarz M, Wolff G, Bird ED, Vonsattel JP, Gusella JF (1993) De novo expansion of a (CAG)n repeat in sporadic Huntington's disease. Nat Genet 5:168–173. [Myers RH, MacDonald ME, Gusella JF (1993) Discrepancy resolved. Nat Genet 5:215]

Nagafuchi S, Yanagisawa H, Sato K, Shirayama T, Ohsaki E, Bundo M, Takeda T, Tadokoro K, Kondo I, Murayama N, Tanaka Y, Kikushima H, Umino K, Kurosawa H, Furukawa T, nihei K, Inoue T, Sano A, Komure O, Takahashi M, Yoshizawa T, Kanazawa I, Yamada M (1994) Dentatorubral and pallidoluysian atrophy expansion of an unstable CAG trinucleotide on chromosome 12p. Nat Genet 6:14–18

Naito H., Oyanagi S (1982) Familial myoclonus epilepsy and choreoathetosis: hereditary dentatorubral-pallidoluysian atrophy. Neurology 32:798–807

Nasir J, Floresco JB, O'Kusky JR, Diewert VM, Richman JM, Zeisler J, Borowski A, Marth JD, Phillips AG, Hayden MR (1995) Targeted disruption of the Huntington's disease gene results in embryonic lethality and behavioral and morphological changes in heterozygotes. Cell 81:811–823

Norremolle A, Riess O, Epplen JT, Fenger K, Hasholt L, Sorensen SA (1993) Trinucleotide repeat elongation in the Huntingtin gene in Huntington disease patients from 71 Danish families. Hum Mol Genet 2:1475–1476

Norremolle A, Sorensen SA, Fenger K, Hasholt L (1995a) Correlation between magnitude of CAG repeat length alterations and length of the paternal repeat in paternally inherited Huntington's disease. Clin Genet 47:113–117

Norremolle A, Nielsen JE, Sorensen SA, Hasholt L (1995b) Elongated CAG repeats of the B37 gene in a Danish family with dentato-rubro-pallido-luysian atrophy. Hum Genet 95:313–318

Novelletto A, Persichetti F, Sabbadini G, Mandich P, Bellone E, Ajmar F, Pergola M, Del Senno L, MacDonald ME, Gusella JF, Frontali M (1994) Analysis of the trinucleotide repeat expansion in Italian families affected with Huntington disease. Hum Mol Genet 3:93–98

Osler W (1908) Historical note on hereditary chorea. Browning W (ed) Neurographs. vol 1. Albert C. Huntington Publishing, Brooklyn pp113–116

Persichetti F, Srinidhi J, Kanaley L, Ge P, Myers RH, D'Arrigo K, Barnes GT, MacDonald ME, Vonsattel JP, Gusella JF, Bird ED (1994) Huntington's disease CAG trinucleotide repeats in pathologically confirmed post-mortem brains. Neurobiol Dis 1:159–166

Persichetti F, Ambrose CM, Ge P, McNeil SM, Srinidhi J, Anderson MA, Jenkins B, Barnes GT, Duyao MP, Kanaley L, Wexler NS, Myers RH, Bird ED, Vonsattel JP, MacDonald ME, Gusella JF (1995) Normal and expanded Huntington's disease alleles produce distinguishable proteins due to translation across the CAG repeat. Mol Med 1:374–383

Persichetti F, Carlee L, Faber PW, McNeil SM, Ambrose CM, Srinidhi J, Anderson MA, Barnes GT, Gusella JF, MacDonald ME (1996) Differential expression of normal and mutant Huntington's disease gene alleles. Neurobiol Dis 3: 183–190

Pulst SM, Nechiporuk A, Nechiporuk T, Gispert S, Chen XN, Lopes-Cendes I, Pearlman S, Starkman S, Orozco-Diaz G, Lunkes A, DeJong P, Rouleau GA, Auburger G, Korenberg JR, Figueroa C, Sahba S (1996) Moderate expansion of a normally biallelic trinucleotide repeat in spinocerebellar ataxia type 2. Nat Genet 14: 269–276

Ranum LP, Chung MY, Banfi S, Bryer A, Schut LJ, Ramesar R, Duvick LA, McCall A, Subramony SH, Goldfarb L, Gomez C, Sandkuijl LA, Orr HT, Zoghbi H (1994) Molecular and clinical correlations in spinocerebellar ataxia type I: evidence for familial effects on the age at onset. Am J Hum Genet 55:244–252

Ranum LP, Lundgren JK, Schut LJ, Ahrens MJ, Perlman S, Aita J, Bird TD, Gomez C, Orr HT (1995) Spinocerebellar ataxia type 1 and Machado-Joseph disease: incidence of CAG expansions among adult-onset ataxia patients from 311 families with dominant, recessive, or sporadic ataxia. Am J Hum Genet 57:603–608

Rubinsztein DC, Barton DE, Davison BCC, Ferguson-Smith MA (1993) Analysis of the huntingtin gene reveals a trinucleotide-length polymorphism in the region of the gene that contains two CCG-rich stretches and a correlation between decreased age of onset of Huntington's disease and CAG repeat number. Hum Mol Genet 2:1713–1715

Rubinsztein DC, Leggo J, Coles R, Almqvist E, Biancalana V, Cassiman JJ, chotai K, Connarty M, Craufurd D, Curtis A, Curtis D, Davidson mJ, Differ AM, Dode C, Dodge A, Frontali M, Ranen NG, Stine OC, Sherr M, Abbott MH, Franz ML, Graham CA, Harper PS, Hedreen JC, Jackson A, Kaplan JC, Losekoot M, MacMillan JC, Morrison P, Trottier Y, Novelletto A, Simpson SA, Theilmann J, Whitaker JL, Folstein SE, Ross CA, Hayden MR (1996) Phenotypic characterization of individuals with 30–40 CAG repeats in the Huntington disease (HD) gene reveals HD cases with 36 repeats and apparently normal elderly individuals with 36–39 repeats. Am J Hum Genet 9:16–22

Sanpei K, Takano H, Igarashi S, Sato T, Oyake M, Sasaki H, Wakisaka A, Tashiro K, Ishida Y, Ikeuchi T, Koide R, Saito M, Sato A, Tanaka T, Hanyu S, Takiyama Y, Nishizawa M, Shimizu N, Nomura Y, Segawa M, Iwabuchi K, Eguchi I, Tanaka H, Takahashi H, Tsuji S (1996) Identification of the spinocerebellar ataxia type 2 gene using a direct identification of repeat expansion and cloning technique, DIRECT. Nat Genet 14:277–284

Sato K, Kashihara K, Okada S, Ikeuchi T, Tsuji S, Shomori T, Morimoto K, Hayabara T (1995) Does homozygosity advance the onset of dentatorubral-pallidoluysian atrophy? Neurology 45:1934–1936

Schmitt I, Baechner D, Megow D, Henklein P, Boulter J, Hameister H, Epplen JT, Riess O (1995) Expression of the Huntington disease gene in rodents: Cloning the rat homologue and evidence for down regulation in non-neuronal tissues during development. Hum Mol Genet 4:1173–1182

Schöls L, Menezes Saecker-Vieira AM, Schöls S, Przuntek H, Epplen JT, Riess O (1995) Trinucleotide expansion within the MJD1 gene presents clinically as spinocerebellar ataxia and occurs most frequently in German SCA patients. Hum Mol Genet 4:1001–1005

Schöls L, Amoiridis G, Langkafel M, Przuntek H, Riess O (1995) Machado-Joseph disease mutations as the genetic basis of most spinocerebellar ataxias in Germany. J Neurol Neurosurg Psychiatry 59:449–450

Servadio A, Koshy B, Armstrong D, Antalffy B, Orr HT, Zoghbi HY (1995) Expression of the ataxin-1 protein in tissues from normal and spinocerebellar ataxia type 1 individuals. Nat Genet 10:94–98

Sharp AH, Loev SJ, Schilling G, Li SH, Li XJ, Bao J, Wagster MV, Kotzuk JA, Steiner JP, Lo A, Hedreen J, Sisodia S, Snyder SH, Dawson TM, Ryugo DK, Ross CA (1995) Widespread expression of Huntington's disease gene (IT15) protein product. Neuron 14:1065–1074

Snell RG, MacMillan JC, Cheadle JP, Fenton I, Lazarou LP, Davies P, MacDonald ME, Gusella JF, Harper PS, Shaw DJ (1993) Relationship between trinucleotide repeat expansion and phenotypic variation in Huntington's disease. Nat Genet 4:393–397

Soong BW, Wang JT (1995) A study on Huntington's disease associated trinucleotide repeat within the Chinese population Proc Natl Sci Counc Repub China B 19:137–14

Stine OC, Pleasant N, Franz ML, Abbott MH, Folstein SE, Ross CA (1993) Correlation between the onset age of Huntington's disease and length of the trinucleotide repeat in IT-15. Hum Mol Genet 2:1547–1549

Stine OC, Li SH, Pleasant N, Wagster MV, Hedreen JC, Ross CA (1995) Expression of the mutant allele of IT-15 (the HD gene) in striatum and cortex of Huntington's disease patients. Hum Mol Genet 4:15–18

Strong TV, Tagle DA, Valdes JM, Elmer LW, Boehm K, Swaroop M, Kaatz KW, Collins FS, Albin RL (1993) Widespread expression of the human and rat Huntington's disease gene in brain and nonneural tissues. Nat Genet 5:259–265

Takiyama Y, Igarashi S, Rogaeva EA, Endo K, Rogaev EI, Tanaka H, Sherrington R, Sanpei K, Liang Y, Saito M, Tsuda T, Takano H, Ikeda M, Lin C. Chi H, Kennedy JL, Lang AE, Wherrett JR, Segawa M, Nomura Y, Yuasa T, Weissenbach J, Yoshida M, Nishizawa M, Kidd KK, Tsuji S, St George-Hyslop PH (1995) Evidence for inter-generational instability in the CAG repeat in the MJD1 gene and for conserved haplotypes at flanking markers amongst Japanese and Caucasian subjects with Machado-Joseph disease. Hum Mol Genet 4:1137–1146

Telenius H, Kremer B, Goldberg YP, Theilmann J, Andrew SE, Zeisler J, Adam S, Greenberg C, Ives EJ, Clarke LA, Hayden MR (1994) Somatic and gonadal mosaicism of the Huntington disease gene CAG repeat in brain and sperm [published erratum appears in Nat Genet 1994;7:113]. Nat Genet 6:409–414

Trottier Y, Biancalana V and Mandel JL (1994) Instability of CAG repeats in Huntington's disease: relation to parental transmission and age of onset. J Med Genet 31:377–382

Trottier Y, Devys D, Imbert G, Saudou F, An I, Lutz Y, Weber C, Agid Y, Hirsch EC, Mandel JL (1995a) Cellular localization of the Huntington's disease protein and discrimination of the normal and mutated form. Nat Genet 10:104–110

Trottier Y, Lutz Y, Stevanin G, Imbert G, Devys D, Cancel G, Saudou F, Weber C, David G, Tora L, Agid Y, Brice A, Mandel J-L (1995b) Polyglutamine expansion as a pathological epitope in Huntington's disease and four dominant cerebellar ataxias. Nature 378:403–406

Vessie PR (1932) On the transmission of Huntington's chorea for 300 years: the Bures family group. J Nerv Ment Dis 76:553–573

Vonsattel J-P, Myers RH, Stevens TJ, Ferrante RJ, Bird ED, Richardson EP Jr (1985) Neuropathological classification of Huntington's disease. J Neuropathol Exp Neurol 44:559–577

Warner TT, Williams LD, Walker RW, Flinter F, Robb SA, Bundey SE, Honavar M, Harding AE (1995) A clinical and molecular genetic study of dentatorubropallidoluysian atrophy in four European families. Ann Neurol 37:452–459

Wexler NS, Young AB, Tanzi RE, Travers H, Starosta-Rubenstein S, Penney JB, Snodgrass SR, Shoulson I, Gomez F, Ramos-Arroyo MA, Penchaszadeh G, Moreno R, Gibbons K, Faryniarz A, Hobbs, W, Anderson MA, Bonilla E, Conneally PM, Gusella JF (1987) Homozygotes for Huntington's disease. Nature 326:194–197

Whitefield JE, Williams L, Snow K, Dixon J, Winship I, Stapleton PM, Faull RM, Love DR (1996) Molecular analysis of the Huntington's disease gene in New Zealand. N Z Med J 109:27–30

Wiggins S, Whyte P, Huggins M, Adam S, Theilmann J, Bloch M, Sheps SB, Schechter MT, Hayden MR (1992) The psychological consequences of predictive testing for Huntington's disease. N Engl J Med 327:1401–1405

Wright TJ, Ricke DO, Denison K, Abmayr S, Cotter PD, Hirschhorn K Keinanen M, McDonald-McGinn D, Somer M, Spinner N, Yang-Feng T Zackai E, Altherr MR (1997) A transcript map of the newly defined 165 kb Wolf-Hirschhorn syndrome critical region Hum Mol Genet 6:317–324

World Federation of Neurology Research Group on Huntington's Disease (1990) Ethical issues policy statement on Huntington's disease molecular genetics predictive test. J Med Genet 27:34–38

Xuereb JH, MacMillan JC, Snell R, Davies P, Harper PS (1996) Neuropathological diagnosis and CAG repeat expansion in Huntington's disease. J Neurol Neurosurg Psychiatry 60:78–81

Yazawa I, Nukina N, Hashida H, Goto J, Yamada M, Kanazawa I (1995) Abnormal gene product identified in hereditary dentorubral-pallidoluysian atrophy (DRPLA) brain. Nat Genet 10:99–103

Zeitlin S, Liu J-P, Chapman DL, Papaioannou VE, Efstratiadis A (1995) Increased apoptosis and early embryonic lethality in mice nullizygous for the Huntington's disease gene homologue. Nat Genet 11:155–162

Zhuchenko O, Bailey J, Bonnen P, Ashizawa T, Stockton DW, Amos C, Dobyns WB, Subramony SH, Zoghbi HY, Lee CC (1997) Autosomal dominant cerebellar ataxia (SCA6) associated with small polyglutamine expansions in the alpha1A-voltage-dependent calcium channel. Nat Genet 15:62–68

Zoghbi HY, Orr HT (1995) Spinocerebellar ataxia type 1. Seminars Cell Biol 6:29–35

Zuhlke C, Riess O, Schroder K, Siedlaczck I, Epplen JT, Engel W, Thies U (1993a) Expansion of the (CAG)n repeat causing Huntington's disease in 352 patients of German origin. Hum Mol Genet 2:1467–1469

Zuhlke C, Riess O, Bockel B, Lange H, Thies U (1993b) Mitotic stability and meiotic variability of the (CAG)n repeat in the Huntington disease gene. Hum Mol Genet 2:2063–2067

Myotonic Dystrophy

J. D. Waring[1] and R. G. Korneluk[1]

1

Introduction

1.1
General Comments

Myotonic dystrophy (DM), also known as Steinert's disease, is the most common adult, heritable neuromuscular disease, with an incidence of approximately 1 in 8000. In certain regions the incidence can reach 1 in 450 due to genetic founder effects (Bouchard et al. 1989). DM is an autosomal dominant, multisystemic disorder, characterized by its highly pleiotropic nature and pronounced variability in severity and age of onset. Symptoms can include myotonia (defined as delayed relaxation after a voluntary muscular contraction), progressive weakness and wasting of skeletal muscle, cardiac conduction disturbances, cataracts and retinal degeneration in the eye, mental retardation and other disturbances (hypersomnia and apathy), testicular atrophy and frontal balding in males, endocrine disturbances, and involvement of smooth muscle as evidenced by colonic obstruction and dysphagia. For convenience of classification, the disease can be separated into three distinct, although overlapping categories: (1) a late-onset adult form, in which cataracts and frontal balding in the absence of muscle disturbance are often observed, (2) the classical type of DM, with onset in early or middle adult life, characterized mainly by myotonia and progressive muscular atrophy, mild cognitive dysfunction, and frequent cardiac involvement, and (3) the often fatal congenital form (cDM), characterized by severe hypotonia and respiratory distress, feeding difficulties, facial dysplegia, and mental retardation. In surviving infants with cDM, status often improves and manifestations of the adult form, such as myotonia, appear later. The congenital form was not generally appreciated as DM until long after the adult disorder had been described (Vanier 1960). The frequency of DM, coupled with its high infant mortality, make this disorder an extremely important hereditary muscle disease in humans.

In 1992, after years of effort, a positional cloning approach culminated in the simultaneous discovery by three groups of an unstable region on chromosome

[1] Solange Gauthier Karsh Laboratory, Children's Hospital of Eastern Ontario, 401 Smyth Road, Ottawa, Ontario K1H 8L1, CANADA

19 as the genetic basis for DM (Aslanidis et al. 1992; Buxton et al. 1992; Harley et al. 1992). A polymorphic region at 19q13.3 was found to be larger in DM patients, and the size of this expansion was correlated with the severity of the disease. These reports were quickly followed by the identification of an unstable $(CTG)_n$ trinucleotide repeat (TNR) as the basis for this expansion (Brook et al. 1992; Fu et al. 1992; Mahadevan et al. 1992). Based on sequence homology, the repeat was localized to the 3' untranslated region (UTR) of a gene predicted to encode a serine/threonine kinase (designated myotonic dystrophy protein kinase, or DMPK). The best matches were originally to vertebrate cAMP-dependent kinases. Prior to the discovery of the DM TNR expansion, a similar basis had been established for spinal and bulbar muscular atrophy (SBMA; Kennedy's disease) and fragile X mental retardation (FRAXA). Subsequently, the family of TNR diseases expanded quickly to include Huntington's disease (HD), spinocerebellar ataxia type I (SCA1), FRAXE mental retardation, dentatorubral pallidoluysian atrophy (DRLPA), Machado-Joseph's disease (MJD; SCA3), and Friedreich's ataxia (FA). More recently, TNR expansion has been found to be the basis for SCA2 (Imbert et al. 1996; Pulst et al. 1996; Sanpei et al. 1996), and expansion at the fragile site FRA11B has also been associated with some cases of Jacobsen syndrome (Jones et al. 1995). (For recent reviews pertaining to information presented on these diseases, please see Ashley and Warren 1995, Timchenko and Caskey 1996, and Warren 1996.)

The TNR involved in these diseases are present on normal chromosomes with a polymorphic copy number, similar to microsatellite sequences, and are transmitted in a stable fashion in successive generations. A common pattern can be discerned for many of the TNR disorders, as they are usually dominantly inherited and have a neurological component. Also, the repeat sequences are generally found within exons, and although the nomenclature occasionally varies depending upon the polarity or initial nucleotide used, most have a CXG sequence: either CGG or CTG. A CGG repeat (coding strand polarity) is present in the 5'-UTR of the FMR-1 and CBL2 genes at the FRAXA and FRA11B loci, respectively, and a CCG in the 5'-UTR of the FMR-2 gene at the FRAXE loci (Ashley and Warren 1995). Apart from DM, a CAG repeat is found within the coding sequences of the respective genes for the remaining disorders. FA has provided a notable exception to this pattern, as it is recessively inherited and is caused by an expanded AGG repeat within the first intron of the frataxin gene (Campuzano et al. 1996). The hallmark of TNR's is their pronounced instability above a certain size threshold, which varies for the different diseases. Mutant repeats exhibit a tendency towards further expansion in successive generations, which depends upon length. The term "dynamic" mutation was coined to indicate this influence of repeat size on further mutability (Richards and Sutherland 1992). Finally, genetic anticipation has been associated with TNR diseases to varying degrees. Anticipation is defined as an increase in the severity of the disease or penetrance in subsequent generations, often accompanied by an earlier age of onset. Although previously dismissed as an artifact of ascertainment bias, anticipation had been validated for DM by careful genealogical studies (Höweler et al. 1989; Ashizawa et al. 1992a). The discovery of TNR expansion has provided a molecular basis for anticipation, as generational size gains can be correlated with increased disease severity and an earlier age of onset. For DM, the CTG repeat copy number varies from approximately 5 to 37 in normal individuals, and from 50 to several thousand repeats in individuals with disease. This degree of repeat instability and the resultant anticipation which can be observed in pedigrees, along with the corresponding variability in disease severity, are more pronounced in DM than in any other TNR disease.

To date, the exact function of the gene affected by expansion of a TNR is known only for SBMA, where the repeat lies within the first exon of the androgen receptor (AR) (La Spada et al. 1991). Truncations close to the FRA11B locus have been associated with a CCG expansion within the CBL2 proto-oncogene, a receptor tyrosine kinase (Jones et al. 1995), although the functional significance of this is not known. The FMR-1 protein associates with ribosomes via RNA (Khandjian et al. 1996; Tamanini et al. 1996) and shows enhanced expression in germ cells during proliferation (Bachner et al. 1993). The other diseases are primarily neurodegenerative in nature and are believed to be caused by a toxic gain of function conferred by a polyglutamine (CAG coding) expansion upon their respective gene products. The substrate(s) for the DMPK protein and the pathway in which it functions have remained elusive. In addition, the position of the repeat expansion within the 3'-UTR has been difficult to reconcile with the dominant mode of inheritance. Expansion of CGG repeats into the disease range at the FRAXA site is associated with abnormal hypermethylation at an adjacent CpG island, which is also highly methylated on the inactivated X chromosome in normal females. This causes disease by transcriptional silencing, and to some degree translational inhibition, in affected males and a proportion of carrier females (depending upon lyonization). In contrast, DMPK expression cannot be reduced below 50 % of the normal level in affected patients unless a position variegation effect operates to reduce the level of the normal allele. It is therefore possible that the CTG expansion effects disease in some other fashion, perhaps related to its presence on mutant transcripts. Clear answers regarding the effects of the TNR expansion on DMPK gene expression have been slow to emerge, primarily because of the difficulty in quantitating the products of the mutant allele.

Identification and cloning of the defect responsible for DM now allows for prenatal screening and intervention. However, because of the high incidence in certain populations and the severity of the congenital form, it continues to be a significant health concern. Although anticipation should eventually cause extinction of the mutation within a family by reduction of reproductive fitness, current thought is that novel expansions will continue to emerge from a pool of individuals harboring CTG expansions at the upper end of the normal range.

In addition, DM continues to pose a significant and interesting challenge to researchers several years after the cloning of the DMPK gene. Considerable effort has been devoted to the study of factors affecting the evolution, inheritance, and instability of the DM CTG repeat. While the mechanism by which this expansion results in myotonic dystrophy is not yet known, important information regarding the pathophysiology has already emerged and continues to do so. In this chapter, we will attempt to review the literature in the major research areas subsequent to the identification of the DM mutation, with an eye toward identifying factors likely to be important in the DM disease process. The basis of the instability of TNR will not be dealt with to any great extent here, as this will be the subject of another chapter. Nevertheless, there are some genetic features shared by the different TNR disorders, as well as some which are unique to DM, which we will attempt to highlight. Before proceeding, we will briefly discuss the major pathological features of DM in order to provide a physiological context for the work referenced herein. Readers are referred to an excellent text by Harper (1989) for the clinical references related to this section and a more thorough description of the manifestations of the disease.

1.2
Clinical Features

As mentioned, DM is highly variable, and no one symptom is constant. Often, the most evident feature of DM is a characteristic facial appearance, a result of general weakness of the superficial muscles and hollowing at the temples as a consequence of jaw muscle wasting. Ptosis, and frontal balding in males are also common features. Generally, muscle involvement is restricted to the face, jaw, neck, and distal limb muscles. Weakness of the sternomastoids and anterior neck muscles is typically marked. Affected muscle has a number of distinguishing features at the histological level. An increase in the number of central nuclei in cross sections of muscle fibers (and in nuclear chains in longtitudinal sections) is seen in many disorders, but is very prominent in DM. Ringed fibers, which have striations arranged circumferentially around the fiber, are seen in approximately 70 % of biopsies. Sarcoplasmic masses, which are abnormal subsarcolemmal structures composed of disorganized myofilaments and aggregated tubules, may also be present. The basis for these defects is not known, but centronucleation is believed to reflect immaturity of a developing fiber. Fiber sizes show increased variation, and there is a pronounced atrophy of type I (slow, oxidative) fibers, sometimes accompanied by hypertrophy of type II fibers. At the ultrastructural level, degenerative changes of myofibrils, particularly at the Z line and I band, are sometimes observed. Muscle spindles show increased splitting, with degenerative changes and increased numbers of intrafusal fibers.

The pathological changes observed in adult muscle are not nearly as common in cDM muscle, the exception being centronucleated fibers. Instead, there is a pronounced hypotonia, affecting both the diaphragm and limb muscles, and the distinctive presence of numerous satellite cells. This profile is consistent with a general delay in muscle development. The fatal nature of cDM is invariably due to respiratory insufficiency.

The hallmark symptom of DM is myotonia, defined as an inability to relax after a voluntary contraction, which is usually clinically evident in symptomatic cases. Electromyographic examination reveals increased insertion activity, and characteristic myotonic runs (prolonged trains of repetitive action potentials gradually declining in amplitude over time) can be elicited by stimulation. "Warm-up," or repetitive muscle activity, decreases both clinical and electrical myotonia. Dystrophic changes such as reduced action potentials can also be seen.

Clinical symptoms associated with cardiac defects are rarer, although congestive heart failure or mitral valve prolapse often result in sudden death. However, abnormalities are easily demonstrated by electrocardiogram, including a variety of conduction defects, arrhythmias, and other changes. These findings are consistent with the selective destruction of the conductive system and myocardium and the accompanying fibrosis and fatty infiltration, which are often seen upon autopsy. Smooth muscle involvement is not often appreciated, but is apparent in many systems. Aspiration pneumonia is a common complication of dysphagia, and hydramnios may also reflect a swallowing defect in fetuses. Dilatation and reduced motility can be demonstrated in other parts of the gastrointestinal tract. Pregnancies may be complicated by uncoordinated contractions during labor, and low intraocular pressure in the eye is suspected to result from a ciliary muscle defect.

The numerous other system defects are striking in their lack of a common discernible physical basis. Apart from muscular involvement, a distinctive cataract is most often associated with DM. These appear as small, multicolored, iridescent granules with a dust like appearance and are present in both the anterior and posterior subcapsular regions. Microscopically, lens epithelia have unusually long extensions of the cytoplasmic membrane into the capsule. Cataracts appear to be formed of vacuoles containing membranous multilamellar whorls. Retinal integrity is occasionally disturbed by a peripheral pigmentary degeneration similar to retinitis pigmentosa, or a central macular degeneration. Testicular atrophy is observed in autopsy and biopsy studies with a high frequency. Changes include degeneration and fibrosis of tubular cells, hyperplasia of interstitial cells, and reduced spermatogenesis. Reduced fertility is commonly noted in pedigrees, but appears to be more due to celibacy than sterility for affected males. Several endocrine abnormalities have also been reported. Abnormal glucose tolerance resulting in hyperinsulinemia can be demonstrated, and there is a small increase in the risk of clinical diabetes. Modest reductions in serum testosterone are sometimes seen, along with increased levels of pituitary gonadotropins believed to be secondary to the gonadal dysfunction.

Hypersecretion of adrenocorticotropic hormone in response to corticotrophin releasing hormone can be demonstrated in patients, although there is no evidence of adrenal dysfunction. This is speculated to be secondary to T-type voltage-gated Ca^{2+} channel defects (Hockings et al. 1993). Severe mental retardation is common in cDM infants, but reduction of IQ in adults is modest or absent. However, accounts of DM patients describe them as being apathetic and inactive, with hypersomnia a common observation. A variety of other changes have been reported, including decreased brain weight, ventricular dilatation, and disorded cortical neurons, which are perhaps indicative of an ongoing degenerative process. There are reports of numerous other related and minor system defects. The unknown relationship of these diverse changes to the more central muscular involvement is perhaps the most unique and intriguing aspect of this disease.

2
Genetics and Transmission

2.1
Chromosome Analysis and Disease Evolution

To date, no case of DM has been attributed to any other mutational defect besides the CTG repeat expansion. Although a number of cases clinically consistent with DM which have normal CTG alleles have been noted (e.g., Thornton et al. 1994a), it can be estimated that expansion accounts for at least 98 % of cases. Genetic homogeneity is also observed at the level of DM chromosome structure. Pronounced linkage disequilibrium of polymorphic markers on DM chromosomes was observed in early studies, a very surprising finding in light of DM's dominant inheritance, high frequency, and low reproductive fitness (estimated at 0.7; Harper 1989). Very old mutations are, of course, more characteristic of autosomal recessive disease. Furthermore, anticipation would act to hasten the loss of DM mutations by increasing the loss of fitness in successive generations. This was exemplified by a longitudinal study combining clinical and molecular data from five generations of a DM family which documented the extinction of a mutant gene (de Die-Smulders et al. 1994). Consequently, DM and other TNR diseases such as FRAXA have become interesting genetic models for study. The chromosomal region around the DMPK gene has been subjected to extensive analysis in order to understand the maintenance of the expansion mutation in the face of its continual elimination.

Both prior and subsequent to the identification of the DMPK gene, analysis of polymorphic markers around the DM region (Fig. 1) revealed varying degrees of linkage disequilibrium on DM chromosomes. Linkage within a large, heterogeneous population was first observed for D19S63 in European families, a locus tightly linked to the DM region (Harley et al. 1991, 1992). This observation was subsequently repeated in various other populations (Cobo et al. 1992; Yamagata et al. 1992; Lavedan et al. 1994; Neville et al. 1994; Goldman et al. 1996). More strikingly, a two allele Alu repeat insertion/deletion polymorphism was characterized, and the insertion allele was found to be in complete association with the DM expansion in Caucasian and Japanese populations (Harley et al. 1992; Yamagata et al. 1992; Mahadevan et al. 1993a,b; Goldman et al. 1996). The normal frequency of the insertion allele is 0.51 in people of European ancestry (Imbert et al. 1993), 0.45 in a Japanese population (Yamagata et al. 1992) and 0.71 in South African negroids (Goldman et al. 1995). Only one exception to this association has been characterized to date, namely a Nigerian male affected with DM (Krahe et al. 1995a). Thus the disease appears to be the result of one or a few ancestral mutations. In addition, more pronounced linkage has been observed for various markers within more discrete populations, indicative of founder effects, such as that found for alleles within the apolipoprotein C2 (APOC2) and APOE genes in a French Canadian DM population (Thibault et al. 1989; MacKenzie et al. 1989), in APOC2 for a Finnish population

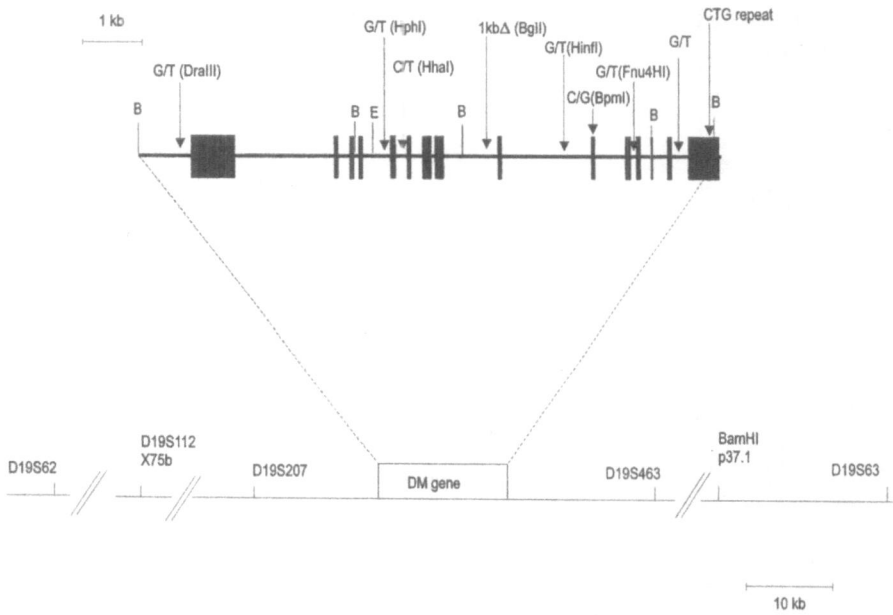

Fig. 1. Sequence polymorphisms at the DM locus. Intragenic (*top*) and extragenic (*bottom*) sequence polymorphisms in linkage disequilibrium with the DM locus. The core (intragenic) haplotype is absolutely conserved on mutant chromosomes (see Table 1), indicating their unique origin. The polymorphism *1kbΔ* is composed of an Alu element in which a 1-kb deletion occurred. The undeleted allele is in total linkage with mutant chromosomes. High linkage values are seen for extragenic markers, with more pronounced scores in various populations due to founder effects. Extragenic polymorphisms are presented as in Goldman et al. (1996; and references therein) with permission; D19S112, D19S62, and D19S63 are approximately 90, 280, and 140 kb, respectively, from the DM gene. Intragenic polymorphisms are presented as in Neville et al. (1994)

(Nokelainen et al. 1990), a *Bam*HI polymorphism at marker p37.1 in a Japanese population (Yamagata et al. 1992), and extragenic markers D19S63 and the CA repeat polymorphism at D19S112 in the South African caucasian DM population (Goldman et al. 1996)

The striking degree of genetic homogeneity for DM chromosomes suggested by linkage analysis has been supported by haplotype studies. The distribution of CTG alleles is modal (Fig. 2), with $(CTG)_5$ being the most common allele, and a second major peak occurring at $(CTG)_{11-14}$ in European populations (Davies et al. 1992). Allele sizes of 19 or longer are rare, accounting for approximately 10 % of the total, with no one greater than 1 %. Imbert et al. (1993) examined a French population for the distribution of CTG alleles in relation to the Alu insertion/deletion and D19S112 polymorphisms. CTG allele sizes of 5 and 19 or greater (including DM chromosomes) were found exclusively associated with the Alu insertion, while $(CTG)_{11-13}$ alleles were observed only on Alu deletion chromosomes. Furthermore, D19S112 showed a similar, marked disequilibrium pattern for $(CTG)_{19-36}$ normal alleles and DM alleles, implying a common origin. Neville et al. (1994) extended this study using a series of intragenic or closely linked markers in a broader population and identified four major and five derivative haplotypes (Table 1). DM was in complete association

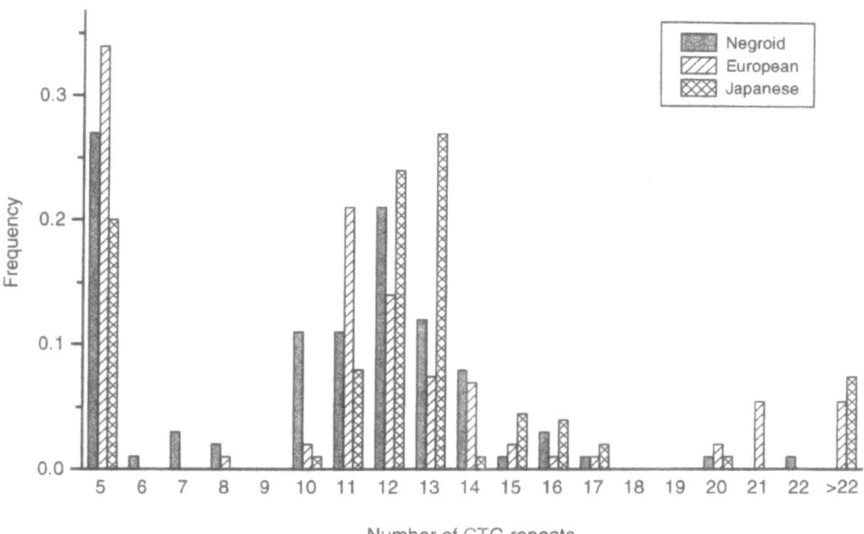

Fig. 2. Distribution of DM protein kinase (DMPK) CTG alleles in three different populations, illustrating the highly polymorphic nature and modal distribution of normal CTG repeats. The most frequent mode at $(CTG)_5$ and $(CTG)_{\geq 19}$ alleles are related by the presence of the Alu insertion allele (Fig. 1); chromosomes with $(CTG)_{11-13}$ alleles contain the deletion allele in most cases. Note the paucity of alleles longer than 14 and the absence of alleles longer than 22 in the negroid population (which has a very low incidence of DM). (From Goldman et al. 1994 and reference to Davies et al. 1994 therein, with permission).

Table 1. Myotonic dystrophy and normal haplotype frequencies (Neville et al. 1994)

Haplotype	Sequence Polymorphisms									Haplotype Frequency (%)
	DMR N9 (G/T) Dra III Dra III	DMK (G/T) HphI	DMK (C/T) HhaI	DMK (Δ1kb)	DMK (G/T) HinfI	DMK (C/G) BpmI	DMK (G/T) Fnu4HI	DMK(G/T)	pCN400 (D19S463) TaqI	
DM	1	2	1	1	2	2	1	T	2	100
A	1	2	1	1	2	2	1	T	2	49
B	2	2	2	2	1	2	2	G	1	27
C	1	1	2	2	1	1	2	G	1	16
D	1	2	2	2	1	2	2	G	1	8
E	2	2	2	2	1	2	2	G	2	<1
F	1	1	2	2	1	2	2	G	1	<1
G	1	2	1	1	2	1	1	T	2	<1
H	2	2	2	2	2	2	2	G	2	<1
I	1	1	2	2	2	2	1	G	1	<1

Haplotypes were constructed from intragenic and one flanking polymorphism (D19S463), in a diverse Canadian DM population. The DM haplotype is absolutely conserved, indicating the unique origin of all mutant chromosomes, and is also the most frequent in the normal population.

with the most common haplotype (haplotype A, including the Alu insertion allele), representing approximately 50 % of the population studied. Using the more distal marker D19S63, a similar pattern of disequilibrium was again seen for haplotype A individuals with $(CTG)_{\geq 19}$ alleles and DM patients. This same general pattern has been seen in other studies (Lavedan et al. 1994; Yamagata et al. 1996). It can therefore be concluded that all cases of DM in these populations arose from a single ancestral chromosome and that the $(CTG)_{19}$ alleles were derived from $(CTG)_5$ by one or a few events. Furthermore, the CTG_{19-36} alleles most likely represent the pool from which novel DM mutations arise.

The incidence of DM is very low in ethnic Africans, Cantonese, Thais and Oceanians (Ashizawa and Epstein 1991). Studies on a South African negroid population have demonstrated a greater haplotype diversity than in caucasoids, and the exclusive association of the Alu deletion with $(CTG)_{11-13}$ was not seen (Goldman et al. 1995). There are also many fewer longer CTG alleles (Fig. 2), consistent with the very low frequency of DM (Goldman et al. 1994). It has therefore been theorized that two founder haplotypes $(CTG_5/Alu^+, CTG_{11-13}/Alu^-)$ from a diverse pool left Africa in the migrations which established the Eurasian populations and that the event which founded the $(CTG)_{19}$ alleles on one of them occurred subsequently, but before the various affected races diverged. The almost exclusive emergence of the DM mutation on Alu insertion chromosomes can likely be explained by the fact that they are older (see below), and therefore DM might emerge on any haplotype, given sufficient time. In fact, the unique case of DM found in a Nigerian family on a Alu deletion background appears to be such a novel mutation and demonstrates that the European DM haplotype is not required for the emergence or maintenance of larger repeat sizes (Krahe et al. 1995a). More recent studies on other non-Caucasian populations have further refined this scheme of evolution (Zerylnick et al. 1995; Deka et al. 1996). $(CTG)_5$ alleles on an Alu deletion background have been found, suggesting that the deletion allele may have originated first on $(CTG)_5$, and not $(CTG)_{11-13}$ chromosomes. In addition, larger alleles (17 and 27) on Alu deletion chromosomes were found, again proving that the Alu insertion haplotype is not necessary for longer repeats and moreover, that a $(CTG)_5$ allele gave rise to $(CTG)_{\geq 19}$ multiple times, and not just once. Based on this information, a slower, stepwise evolution of repeats to the larger, at-risk alleles has been proposed to be ongoing (Chakraborty et al. 1996).

The structure of DM chromosomes can be instructively compared to that for other TNR disorders to reveal variations in disease evolution. No linkage disequilibrium has been found for polymorphic markers at the AR locus on SBMA chromosomes, suggesting several original mutations (La Spada et al. 1991). The distribution of normal CAG alleles is 11-31 (La Spada et al. 1992; Edwards et al. 1992), with a mean size of 22 (Fig. 3). No alleles in the size range between those found on normal and SBMA chromosomes (≥ 40) has been reported. SBMA therefore acts more like a traditional X-linked recessive disease, lacking evidence of a unique founding mutation.

In contrast, selected markers on HD and FRAXA mutant chromosomes exhibit disequilibrium. For HD, there are many disease haplotypes, but two major forms account for about 48 % (MacDonald et al. 1992). For FRAXA, a few haplotypes are again preferentially associated with disease (Richards et al. 1992; Oudet et al. 1993). In both cases, these haplotypes are associated with upper

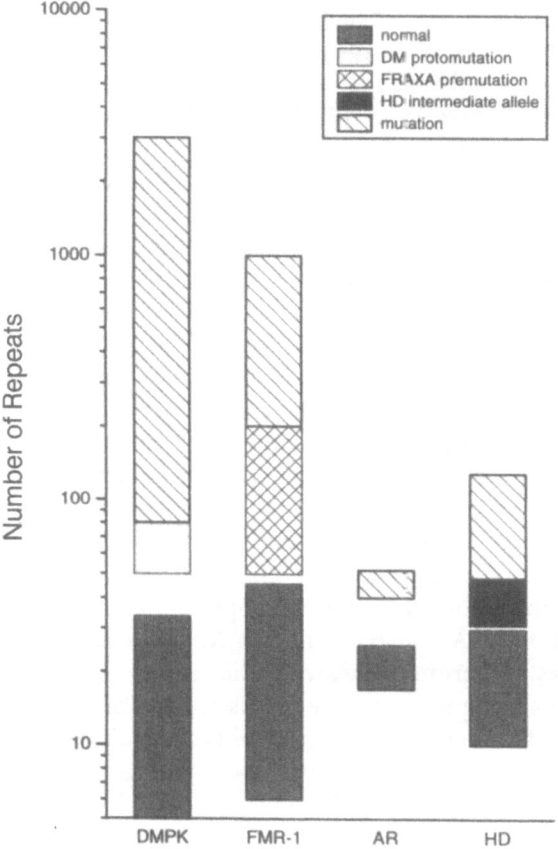

Fig. 3. Distribution of trinucleotide repeat alleles for four different trinucleotide repeat diseases, comparing the repeat distributions for normal, disease, and intermediate classifications for myotonic dystrophy (DM protein kinase, *DMPK*), fragile X mental retardation (FRAXA; *FMR-1*), spinal and bulbar muscular atrophy (SBMA; androgen receptor, *AR*), and Huntington's disease (*HD*). For DM and FRAXA, alleles greater than approximately 50 repeats are committed to eventual expansion to apparent disease alleles. For DM, 50–80 CTG repeats have incomplete penetrance and have been designated protomutations; FRAXA CGG repeats do not manifest disease until >200 and have been designated premutations. The normal range of FRAXA repeats lies close to the committed premutations, while alleles between 37 and 50 for DM are very rare. CAG alleles have not been seen between the normal and disease ranges for SBMA, while HD alleles with incomplete penetrance, termed intermediate alleles (30–37 repeats), account for *de novo* mutations. This may be related to the apparent higher mutability of similarly sized CAG repeats in HD or spinocerebellar ataxia type 2 (SCA2), for instance, as compared with SBMA or SCA1

normal range repeat alleles. Further investigation of HD chromosomes has found that shorter (seven-repeat) alleles at a polymorphic CCG tract adjacent to the CAG repeats (Rubinsztein et al. 1993; Andrew et al. 1994) and an amino acid deletion polymorphism (Δ2634) are strongly associated with longer normal CAG repeats and HD chromosomes (Squitieri et al. 1994; Almqvist et al. 1995; Rubinsztein et al. 1995). Furthermore, rare, meiotically unstable CAG alleles with a size of approximately 30–38 repeats, termed intermediate alleles (IA), were identified on this core haplotype. These were the source of HD in families with a de novo mutation and have subsequently been documented in a number of transmissions (Goldberg et al. 1995). The majority of normal FRAXA CGG alleles are interrupted by AGG triplets, whereas longer, perfect or uninterrupted repeats are those which evolve to mutant alleles (Kunst and Warren 1994) and show the highest disequilibrium with disease haplotypes. HD and FRAXA are therefore similar to DM in that disease chromosomes have arisen from a restricted pool of normal haplotypes, but this association is not as pronounced as for DM.

The idea that disease chromosomes emerge more frequently from the general population for HD and FRAXA than for DM is borne out by comparison of normal allele distributions. An instability threshold for TNR ranging from 35 to 50 repeats is apparent from transmission data, after which mutation rates increase dramatically (see below). Normal HD alleles range from ten to 37 repeats, with a median of 19, while normal FRAXA repeats are distributed with a major mode around 29 and an additional minor mode around 37 (Fu et al. 1991). Therefore, while the major DM repeat allele modes (CTG_5, CTG_{11-14}) appear immune to expansion, much more of the normal populations are close to the instability thresholds for FRAXA and HD. The FRAXA 29 repeat mode probably represents the ancestral chromosome, which has had time to come into equilibrium with many flanking markers (see discussion by Imbert et al. 1993). An overlap can be seen between the upper normal range and the lower end of the mutation range, with repeat sizes as small as 43 observed to be unstable in some transmissions, and repeat sizes of 50–54 transmitted in a stable fashion in several meioses (Fu et al. 1991; Snow et al. 1993; Reiss et al. 1994). Similarly, HD IA class chromosomes are believed to represent an unstable transition state from long normal alleles with incomplete penetrance. In contrast, alleles in the transition class (30–50 repeats) are rare for DM. Thirty-five and 42 repeats have been reported in patients diagnosed with DM (Brunner et al. 1993a; Redman et al. 1993). More recently, a 42-CTG repeat was found to expand over two generations into a full mutation in a Japanese family (Yamagata et al. 1994).

2.2
Predisposing Haplotypes and Mutation Rates

The consequences of repeat expansion to the instability threshold varies. Repeats in the range of 50–100 result in no or very minimal disease for FRAXA and DM, whereas repeats above $(CAG)_{40}$ are often fully symptomatic for disease with coding region expansions (Fig. 3). An asymptomatic FRAXA premutation state has been defined for carrier females and normal transmitting males, with CGG repeat sizes of approximately 50–200. For DM, alleles in the 50- to 80 CTG repeat range were designated "protomutations," as patients will sometimes present with mild symptoms, generally cataracts at an advanced age (Barceló et al. 1993). DM and FRAXA repeats can reach several thousands in length, while large CAG expansions (>100) are rare, suggesting that this would result in a dominant lethal inactivation of the respective polyglutamine-containing protein products.

As implied above, certain haplotypes are at risk for disease either because they carry higher normal alleles more prone to expansion or because these chromosomes contain elements which confer greater mutation rates on the triplet repeats in *cis*. Many authors have emphasized the former possibility. Measurements of mutation rates have demonstrated increasing mutability with size for simple dinucleotide repeats (Weber 1990). The finding that long, perfect, normal repeats are those that evolve to disease alleles for FRAXA (above), and also for SCA1, SCA2, and MJD (Orr et al. 1993; Kawaguchi et al. 1994; Imbert et al. 1996; Pulst et al. 1996; Sanpei et al. 1996), suggests that the loss of stabilizing interruptions is the key to further expansion. This is consistent with the notion that the length of the repeat alone is the driving force behind expansion. A single case has been reported of an unusual imperfect DM 37 repeat which appears to have evolved as a (CCGCTG) hexamer and exhibits greater stability than a perfect 27 repeat allele (Leeflang and Arnheim 1995).

For DM, the Alu insertion allele within the gene is the ancestral one, based on both its structure and its presence in primates (Mahadevan et al. 1993b; Rubinsztein et al. 1994). The association of DM with the more ancient haplotype (with one exception) therefore may be a consequence of the extended time available for the emergence of longer normal CTG alleles. Imbert et al. (1993) proposed that the unique or rare events which converted the ancestral $(CTG)_5$ to $(CTG)_{19}$ initiated a gradual drift to larger sizes. Infrequent conversion of these larger, at-risk alleles into the 30–50 range would greatly increase the likelihood of further mutation to $(CTG)_{>50}$, which then commits a family to eventual disease manifestation. CTG alleles in the 30–50 range were estimated to occur with a frequency of $1–2\times10^{-3}$, while the frequency of alleles >50 was estimated to be approximately 10^{-4}. The mutation rate for long normal alleles was placed between that for (CA) repeat microsatellites and an estimated rate for SBMA normal alleles (which have a very similar distribution to the $(CTG)_{19-36}$

mode; Edwards et al. 1992), i.e., between 10^{-3} and 10^{-4}. Assuming a state of equilibrium, the replacement of eliminated mutant chromosomes suggests an efficient conversion of $(CTG)_{30-50}$ to DM alleles, with a rate ranging from 0.01 to 0.5 (with increasing length).

Large population screens which would allow a measure of the rates at which these processes occur during transmission are not available for DM. However, an estimate of the rates at which TNR mutate can be made for males by single sperm typing using small pool (SP)-PCR. At the SBMA locus, $(CAG)_{28-31}$ alleles had an approximately 2.5-fold greater mutation rate than average-sized $(CAG)_{20-22}$ alleles (3.19 % versus 1.31 %; Zhang et al. 1994). However, contractions greatly outnumbered expansions for the larger class (2.9 % and 0.31 %, respectively) and primarily accounted for the increased mutability over the average size class (0.9 % and 0.43 %, respectively). In contrast, sperm from a patient (47 repeats) exhibited 66 % expansions and 15 % contractions (Zhang et al. 1995). Thus, in agreement with the limited transmission data available, an change in repeat number from 20 to 47 increased the contraction rate approximately nine fold, but the rate of expansion increased more than 200-fold. Similarly, average sized alleles at the HD locus (15–18) exhibited only contractions (0.6 %), while larger alleles began to show progressively increasing expansion rates (Leeflang et al. 1995). A $(CAG)_{30}$ allele (near the boundary of normal and mutant alleles) showed 9 % expansion and 3 % contraction rates; an IA allele (36 repeats) had 42 % expansion and 11 % contraction rates; and disease alleles (36–51) had expansion rates increasing to 97 % and contraction rates decreasing to 2 %. Preliminary data for normal DMPK CTG alleles from two donors exhibited the same general pattern (Zhang et al. 1994). Although this data is limited by its restriction to males and will not account for any effects of repeat length on fertility or viability, it is also consistent with the concept that the length of the repeat is the driving force for expansion. Furthermore, these results suggest a fundamental distinction between expansion and contraction, with expansion being relatively rare until a threshold is reached.

Once repeats are in the pre- or protomutation range, a strong relationship between repeat size and intergenerational expansion of the repeat is apparent. For FRAXA, the presence of 90 or more CGG repeats resulted in expansion in 100 % of cases (Fu et al. 1991; Heitz et al. 1992). A very similar degree of instability, with commitment to expansion for $(CTG)_{>80}$, was observed for DM (Barceló et al. 1993; Brunner et al. 1993a; Harley et al. 1993). Intergenerational increases of DM protomutations are much more frequent than transmissions without a change, although stable transmission of protomutations are not as likely to be seen after ascertainment of a clinically affected kindred due to anticipation and may therefore still contribute to maintenance of the disease in the population.

However, certain evidence has been noted in favor of the existence of *cis* elements predisposing haplotypes to expansion. For example, a positive correlation between the degree of mosaicism of HD CAG repeats in sperm (as compared with blood) and the likelihood of expansion during transmission was seen (Telenius et al. 1995). Reductions were seen where mosaicism was most limited. Likewise, a DM patient who transmitted an apparent reduction to two of his offspring had a much smaller and more limited distribution of allele sizes in sperm compared with blood (Monckton et al. 1995). While greater postzygotic expansion in blood than in sperm may have masked the fact that small expansions were actually transmitted in this case (see below), there appears to be a relationship between the degree of mosaicism which develops in sperm and the likelihood of transmitting a large expansion. Also, expansion rates estimated by SP-PCR for normal alleles of similar sizes vary greatly for SBMA (28–31 CAG repeats, 0.43 %), DM (27 CTG repeats, 6 %), and HD (30 CAG repeats, 9 %). Mosaicism in sperm of SBMA patients is lower than that in the other $(CAG)_n$ disorders, and anticipation is correspondingly rarer (Watanabe et al. 1996). There may be more to learn regarding chromosomal elements which augment or suppress instability from normal alleles at these different loci.

2.3
Segregation Distortion

Some authors have sought further mechanisms which could explain the prevalence of DM. The rapid removal of expanded alleles from pedigrees and the observed size reductions seen for normal alleles prompted the proposal that the preferential transmission of longer alleles would be required to explain the maintenance of the at risk $(CTG)_{\geq 19}$ pool. Carey et al. (1994) examined meioses from normal heterozygotes with one allele with a repeat size of ≥19 and found that the larger allele was transmitted in 56.4 % of cases overall, with a significant distortion for paternal transmissions. Genarelli et al. (1995) documented transmissions from affected individuals and found that the offspring inherited the mutant chromosome in 58.1 % of cases. Again, this distortion was more evident for paternal transmissions and was primarily accounted for by transmissions to sons. A case was made for an increase in the DM gene pool via father-to-son transmission of (proto)mutations. However, Hurst et al. (1995) reanalyzed the results of these two studies and concluded that the findings of preferential male-specific transmission of longer alleles was not substantiated by the data. These studies were followed by two in which significant female-specific segregation distortion was seen: Shaw et al. (1995) examined transmissions from normal heterozygotes and observed 57.7 % inheritance of the longer allele from mothers, while Chakraborty et al. (1996) likewise found that transmission of the longer allele from mothers occurred in 56.5 % of cases. Meiotic drive has been documented in a number of organisms, usually operating at the level of gametogenesis. Typically, sperm carrying a susceptibility allele at a responder locus are inactivated in *trans* by a product with distorter activity from another locus (Lyttle 1993). Leeflang et al. (1996) carefully studied sperm samples from normal heterozygotes for evidence of meiotic drive, taking into account possible effects of contamination and preferential amplification of different allele sizes. They saw no evidence for distortion from an equal ratio of longer and shorter alleles in sperm. They also pointed out that, while the overall findings for distortion in previous studies were significant, the sex-specific differences in transmissions were not. Monckton et al. (1995) also obtained no data in their study consistent with the preferential production of longer alleles in sperm. Male-specific segregation distortion has been noted in pedigrees for other TNR disorders (Ikeuchi et al. 1996). However, further studies will be required to clarify the sex subject to meiotic drive or to identify a postfertilization mechanism through which segregation distortion is operating, in support of the idea that a pool of longer normal CTG repeats is maintained by preferential transmission for DM.

2.4
Trinucleotide Repeat Mosaicism

It was noted early in the genetic study of DM that expansion-containing genomic fragments from peripheral blood tissue had a diffuse or smeared appearance. Heterogeneity consistent with somatic mosaicism had previously been observed for FRAXA, although the patterns were typically composed of more discrete bands, with premutations and full mutations coexisting. Heterogeneity within, and differences in average repeat lengths between, tissues in the same individual was subsequently confirmed for DM (Fig. 4; Ashizawa et al. 1993; Lavedan et al. 1993a; Jansen et al. 1994; Thornton et al. 1994b; Kinoshita et al. 1996). Genotype-phenotype correlations have actively been sought for prognosis and prenatal testing decisions, but the classification of DM into different cat-

MUSCLE
HEART
OVARY
UTERUS
SKIN
LIVER
KIDNEY
FR. CORTEX
THALAMUS
LUNG
THYMUS
ADRENAL
PANCREAS
MES. LYMPHN.
SPLEEN

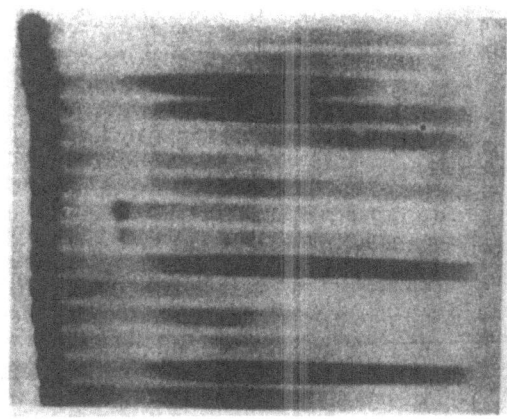

Fig. 4. Somatic heterogeneity. Mutation analysis in 15 different tissues from a DM patient with classical symptoms. Samples were resolved by nondenaturing agarose gel electrophoresis and analyzed by PCR/Southern blotting. Note the differences in intensity and extent of smearing of the larger (mutant) allele in different tissues. The related frontal cortex and thalamus have a distinctly smaller band. (Jansen et al. 1994, with permission.)

egories of severity shows broad overlaps for expansion sizes in blood tissue. It was therefore hoped that better correlations between clinical outcome and the degree of expansion in affected tissues might be observed. As noted by Thornton et al. (1994b), "the idea that muscle disease relates more directly to the CTG repeat amplification in muscle than leukocytes has intrinsic appeal." Surprisingly, expansion status in muscle appears to have little prognostic value. Muscle expansions are typically much larger than those in blood, sometimes up to 13-fold (Anvret et al. 1993; Ashizawa et al. 1993; Thornton et al. 1994b; Zatz et al. 1995). While a weak correlation between the size of the expansion in muscle and in lymphocytes has sometimes been seen (Anvret et al. 1993; Thornton et al. 1994b), no connection between the size of muscle expansions and disease severity or progression is apparent (Anvret et al. 1993; Ashizawa et al. 1993; Zatz et al. 1995). Larger expansions in various tissues may be related broadly to the tissues affected by DM (Kinoshita et al. 1996), but are clearly not sufficient for disease progression. For example, all skeletal muscle expansions were found to be generally large, with severely affected distal muscles not different from unaffected proximal ones (Ashizawa et al. 1993; Thornton et al. 1994b). Overall, cardiac muscle expansions are often found to be the largest for the tissue studied (e.g. Jansen et al. 1994; Thornton et al. 1994b), but many tissue expansions appear to be larger than those in blood. There has been some suggestion of a tissue-specific input upon expansion, as the smallest repeat size was found in the cerebellum not only for DM, but also for the CAG expansions in HD (Telenius et al. 1994), and in MJD and DRLPA (Kinoshita et al. 1996).

When somatic repeat instability occurs and whether it alone can account for the features of intergenerational expansion resulting in anticipation are interesting questions. Regarding the first, early studies did not document heterogeneity before 20 weeks of development. No variations were seen in tissues of fetuses less than 13 weeks old (Hecht et al. 1993; Norbury et al. 1993; Jansen et al. 1994) or in a cDM neonate (Ashizawa et al. 1993), although it may be that in some of these cases the expansions were very large, at the limit of resolution for the detection of size differences. Only small variations were seen in different tissues of a 20 week old fetus (Lavedan et al. 1993a) and in cDM neonates (Jansen et al. 1994; Sabourin et al. 1993). However, in a more recent study, Wöhrle et al. (1995) convincingly demonstrated distinctions in different tissues in a 16 week old fetus (with expansion sizes between 4.8 and 7.8 kb), but not in a 13-week-old fetus. Lines were established from the dura mater of the 16 week old fetus and from myoblasts of the 13 week old fetus (Wöhrle et al. 1995), and clones derived from these lines segregated discrete alleles. Some clones had a repeat size slightly larger than the parental line. Moreover, these lines showed significant expansion over time in culture, even though no tissue differences were seen for the 13 week old fetus. A sigmoidal pattern of size gain was seen, with an apparent lag phase and a plateau. FRAXA repeats exhibited strikingly different behavior in the same experiments. A mosaic pattern of discrete bands was seen in fetal tissues, but this pattern did not vary between tissues (Wöhrle et al. 1993). Fibroblast lines cloned from a fully affected FRAXA male fetus again segregated unique CGG alleles, but the lines were stable in culture. These studies indicate that both repeats likely expand very early after conception, but CGG repeats become fixed while CTG repeats continue to be unstable. The authors suggest that this is due to the extended absence of methylation present around DM expansions after early embryogenesis, while CpG dinucleotides become methylated in the presence of a full mutation FRAXA repeat. This would allow the mismatch repair system to distinguish the parental from daughter strand if slippage occurred during replication. A mouse hypervariable minisatellite with the sequence GGCA also exhibited instability restricted to a very early period after conception (Gibbs et al. 1993).

Consistent with this proposal, it is now apparent that heterogeneity continues to increase postnatally for DM, at least in blood tissue. Mutant alleles have been seen to increase in size in lymphocytes sampled over time, and the gain can be correlated with the initial size of the repeat (Martorell et al. 1995; Wong et al. 1995). It has also been a common observation that genomic fragments from a younger subject are limited in their heterogeneity, even for large expansions. Within the same age category, the degree of heterogeneity can again be correlated with the size of the expansion (Wong et al. 1995). Adult monozygotic twins can develop clearly distinct expansion sizes in their lymphocytes (López de Munain et al. 1994). In apparent contrast, muscle expansions do not appear to enlarge over time in adults (Anvret et al. 1993; Thornton et al. 1994b) and appear to be the same size in different muscle types (Thornton et al. 1994b; Kinoshita et al. 1996). However, it has been noted that the difference in size between muscle and lymphocyte expansions is smaller in younger patients (Zatz et al. 1995) and increases with age. This suggests that muscle expansions grow rapidly postnatally and reach a plateau in early adulthood, consistent with the sigmoidal size gain described above. SP-PCR allows dissection of the smears seen on Southern blot analysis and demonstrates that they are, in fact, composed of multiple unresolved alleles for DM (Fig. 5; Monckton et al. 1995; Wong et al. 1995). It was further verified that blood alleles increase in the range of,

Fig. 5 A, B. Small-pool (SP)-PCR analysis of CTG repeat distribution in sperm and blood of DM ▶ patients. A SP-PCR analysis of blood DNA from one DM individual (DM-2 in **B**). Several reactions for three DNA dilutions containing 1, 4, or 40 amplifiable expanded molecules are shown. Repeat size is shown on the scale at *left*. Blood allele distributions generally exhibit sharp lower boundaries, which for small expansions may represent the size of the inherited allele, and skew towards larger repeats, suggesting highly directional mutation. In contrast, sperm distributions (not shown) are sometimes two-tailed and contain revertant alleles in the high normal range. These data also suggest that expansions are a result of the stepwise gain of small repeat increments. **B** Allele distributions in the sperm (*shaded*) and blood (*black*) of three DM patients by grouping of alleles into 20 repeat size classes. Patients DM-1 and DM-2 exhibit more normal distributions in sperm with an increased average allele size over blood. In contrast, patient DM-3 exhibits a broader pattern and an increased average allele size in blood, consistent with an apparent contraction transmitted to two offspring. However, blood distributions of these offspring (not shown) are actually slightly increased over their father's sperm, suggesting that a small expansion was transmitted. This data also demonstrates that expansion can proceed at different rates in different tissues. (Monckton et al. 1995, with permission)

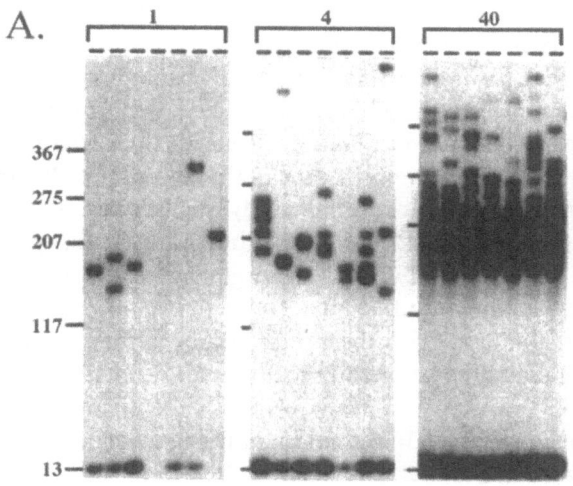

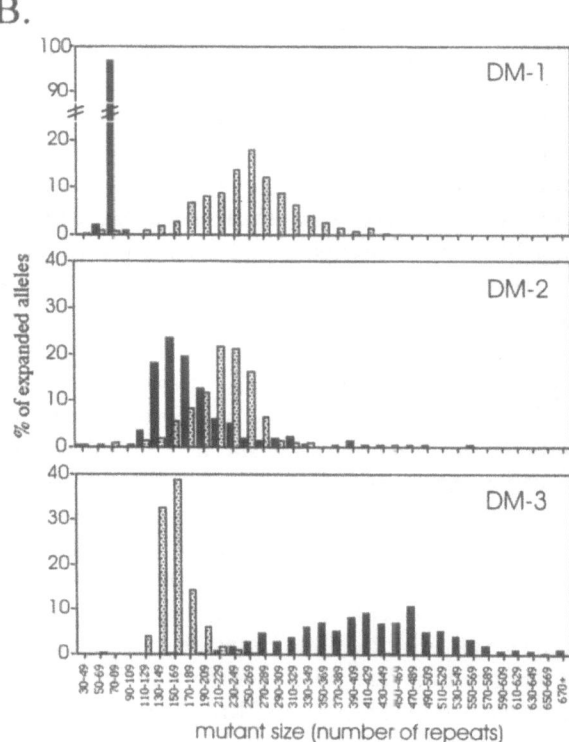

Fig. 5 A, B.

mean, and modal allele size over time (Wong et al. 1995). Results from attempts to replicate mitotic instability by comparing fresh and cultured Epstein-Barr virus-transformed lymphocytes from adult patients have been equivocal. In one study, small differences in average repeat size were seen, with the transformed cells having the larger repeat in most, but not all cases (Ashizawa et al. 1993). Due to overlaps in distributions, however, the authors were not able to rule out clonal selection during culture. In another study, no differences were seen (Lavedan et al. 1993a). Therefore, it is possible that full mutational expansion can only occur in the context of a differentiating organism. Alternatively, these cells may have been near a maximum (or plateau) repeat size. However, these studies clearly indicate that DM CTG repeats are unique in the extent of their dynamic behavior, although the relationship of this extreme instability to disease is uncertain.

Heterogeneity has also been observed in the sperm of DM patients, and full mutations can be demonstrated by Southern blot analysis. The term "gonosomal mosaics" was coined to indicate this combined somatic and germline heterogeneity (Jansen et al. 1994). Can the molecular phenomena underlying anticipation be appreciated strictly in terms of instability occurring postzygotically? No rearrangement of flanking markers is seen upon transmission, so unequal meiotic recombination can be excluded as a mechanism for expansion. It was found that repeat distributions in sperm often had greater variability, and were usually larger on average, than in the blood of the same individual (Jansen et al. 1994). However, sperm distributions were often smaller on average than in the blood of an offspring (consistent with anticipation). PCR analysis of sperm revealed a normal range reversion which had been transmitted to an offspring in one case (Giordano et al. 1994), but this was not seen in two other cases (Brunner et al. 1993b). Based on this low degree of overlap, it was suggested that it was unlikely that a rare sperm with a large expansion (or contraction) was the source of repeat distributions detected in the blood of offspring (Jansen et al. 1994; Wieringa 1994). However, SP-PCR analysis of sperm revealed, in some samples, an increased mean size relative to blood and skewed distribution patterns, interpreted to be consistent with an expansion bias during spermatogenesis (Monckton et al. 1995). This data suggests that passage of an expanded repeat through the male germline is heavily biased towards further expansion. In addition, overlaps in the distribution of parental sperm alleles with blood alleles in some offspring were seen. Therefore, anticipation may be a result of combined size increases from both germline and postzygotic expansion. In contrast, only premutations were detected in sperm from FRAXA males (Reyniers et al. 1993). Two possible explanations for this restriction are the following: (1) expansion from the premutation to the full mutation occurs exclusively during female meiosis, possibly due to epigenetic phenomena, and is followed by regression to the premutation in male germ cells, or (2) expansion is exclusively postzygotic and restricted in developing male germ cells. There may be a selective advantage for sperm with small repeats, perhaps related to the high expression of FMR-1 during gametogenesis (Bachner et al. 1993).

A model has been proposed by Zheng et al. (1993) which accounts for many features of dominant TNR disorders manifesting anticipation. Mitotic instability is viewed as primarily a stochastic process and sufficient to account for variable phenotype and penetrance. Here, the percentage of any tissue which contains a mutation-sized repeat, and therefore the likelihood that symptoms are manifested, depends upon how early in that tissue's development an expansion occurs exceeding the threshold required for symptoms. The point in development at which the expansions reach a threshold may be crucial, although not apparent until years later. Somatic and germline expansion can be viewed as elements of the same process (see also Wieringa 1994). Although this model does not take into account restricted time frames for expansion, as for FRAXA repeats (above), selection against large expansions during spermatogenesis or tissue-specific inputs on expansion, a unified process of mitotic instability does account for major features of DM. Predictions of this model are that (1) anticipation would be more pronounced through male meiosis because of the greater number of divisions, thereby increasing the chance of expansion, and (2) a female giving birth to a severely affected child has an increased chance of her next DM child being severely affected, because the first birth establishes that the percentage of her oocytes containing sufficiently large mutations must be high.

2.5
Effect of Sex on Transmission

Numerous studies following the identification of the DM mutation verified that greatly enhanced instability of repeats above the protomutation threshold was the basis for anticipation observed in pedigrees and that the degree of expansion was positively correlated with the severity of disease and negatively correlated with the age of onset (Ashizawa et al. 1992b; Harley et al. 1992,1993; Hunter et al. 1992; Shelbourne et al. 1992; Tsilfidis et al. 1992; Brunner et al. 1993a; Lavedan et al. 1993a; Redman et al. 1993). Examination of pedigrees also revealed sex-specific influences in transmission patterns. No sex specificity is seen in the overall rate of expansion transmission for DM, while expansion from the premutation to the symptomatic full mutation for FRAXA occurs only during maternal transmissions, the basis for the Sherman paradox. However, it had been noted for some time that the grandparents' generation of affected DM patients more often involves male carriers (Bell 1947). In other words, transmission from the first asymptomatic generation to an affected offspring is predominately paternal. Close inspection of transmission data reveals that expansions of small repeats (less than 100) are more pronounced when paternally transmitted (Brunner et al. 1993a; Wieringa 1994; López de Munain et al. 1995). There is a corresponding paucity of smaller alleles in the protomutation range for females; in some studies, repeats of less than 60 were very rare (Barceló et al. 1993; Brunner et al. 1993a; Harley et al. 1993; Lavedan et al. 1993a). However, as paternal alleles increase in size, the magnitude of expansion diminishes. A negative correlation between paternal repeat size and the magnitude of intergenerational increases was demonstrated in many studies, while no such correlation was seen for maternal transmissions (Cobo et al. 1993; Lavedan et al. 1993a,b; Mulley et al. 1993; Redman et al. 1993; Ashizawa et al. 1994a). Moreover, apparent repeat contractions begin to appear in transmissions from long alleles. In a few rare cases, reversions into the normal range were seen (see below). One key study summarized data from many centers and found that 6.4 % of parent-child transmissions involved contractions (assessed using leukocyte DNA), the majority of which were paternally transmitted (Ashizawa et al. 1994b). This limited ability to expand results in a maximum repeat size of 2000-3000 from males transmissions, and contractions become predominant over 2000 repeats.

However, anticipation was still apparent in approximately half of these reported contractions (Ashizawa et al. 1994b), contrary to expectation. This was much more pronounced for maternal transmissions, and in two cases gave rise to cDM (Cobo et al. 1993). It was suggested that problems of ascertainment may give a false notion of the degree of anticipation occurring, but were unlikely to be entirely accountable. An unusually high number of siblings with contractions was also noted. Comparison of the heterogeneity in germline and somatic tissues by SP-PCR has helped explain these findings (Monckton et al. 1995). Leukocyte distributions typically have a sharp lower boundary, which the authors suggest represents the inherited allele size (Fig. 5). Distributions were skewed towards larger alleles, suggesting a highly directional process. In contrast, sperm exhibited a more normal distribution in two of three cases, with broad tails at both ends. The average allele size was increased relative to blood, consistent with anticipation, but rare alleles within the high normal range could be detected. No such reversions into the normal range were noted for blood, suggesting that this process is specific to germ cells. Surprisingly, a third case exhibited a somewhat opposite pattern, in which the sperm distribution showed relatively little variation, had a sharp lower boundary, and was smaller than that seen in blood. These patterns suggest that expansion from a progenitor allele proceeds at different rates between tissues and individuals and that allele distributions broaden out and becomes more normal as expansion proceeds. The latter patient transmitted an apparent contraction to two of his children, but in fact comparison of the father's sperm distribution with that of his children's blood suggests a small expansion was transmitted. This was presumably masked by the fact that continued somatic expansion had occurred in blood over the lifetime of the father. Together, these results suggest that many contractions are apparent, and not real, consistent with anticipation occurring in approximately half of them. In addition, the lower rate of apparent contractions in female transmissions, coupled with the higher rate of anticipation in these cases, suggest that true contractions are even rarer.

Transmissions which result in cDM cases are almost exclusively maternal. Many observations suggest that the absence of constraint on maternal transmission as compared with paternal transmission explains this exclusivity. First, the average expansion from maternal transmissions is generally greater than from paternal ones (Lavedan et al. 1993a; Harley et al. 1993; Redman et al. 1993). This may be primarily accounted for by transmissions to cDM patients (Tsilfidis et al. 1992; Redman et al. 1993). In another study, this difference disappeared if increases were expressed as a proportion of the parental allele (Harley et al. 1993), although a larger mean size of CTG repeats for females within transmitting and manifesting individuals might explain this latter result (Harley et al. 1993; Lavedan et al. 1993; Ashizawa et al. 1994a). Second, intergenerational increases from longer CTG alleles are more frequent during maternal transmissions (Harley et al. 1993; Redman et al. 1993), and expansions which result in cDM are often more significant than expansions which do not (Lavedan et al. 1993; Redman et al. 1993; Cobo et al. 1995). Third, mothers of cDM offspring have larger repeats on average than mothers of non-cDM offspring (Tsilfidis et al. 1992; Harley et al. 1993; Lavedan et al. 1993; Cobo et al. 1995). Furthermore, maternal repeats greater than $(CTG)_{100}$ were at more than 90 % risk for very large expansions (>400 repeats) in one study (Redman et al. 1993). In summary, two major sex-related dynamics create a typical scenario in which disease first emerges in a family during a male transmission, but the final stages of anticipation which ultimately result in cDM are a consequence of female transmissions.

As referred to above, the initial instability of protomutations in males can be predicted from a stochastic model of mitotic expansion. Spermatogenesis requires many more cell divisions (from 50 to several hundred) than oogenesis (approximately 30), presumably increasing the chance of expansion of small repeats through male transmission. While this effect would become more negligible as average repeat sizes become longer, and further expansion more inevitable, anticipation should be more pronounced through male transmissions overall, in apparent contrast to the maternal specificity for cDM. The negative selection operating on sperm when repeat sizes become longer explains this discrepancy. Molecular evidence for this restriction was obtained by the study of subjects with longer repeat sizes. Sperm distributions often had a lower average size than in blood for these individuals, with a maximum observed size of approximately 1000 repeats (Jansen et al. 1994). In addition, no alleles greater than approximately 1000 repeats were seen in sperm by SP-PCR (Monckton et al. 1995). As previously noted, a more strict restriction operates for FRAXA, but it is not certain whether this occurs at the level of gametogenesis. The mechanistic basis for the reductions which are presumably selected for as a result of this restriction is not known. In one case, haplotype analysis demonstrated a gene conversion event which substituted the mutant allele with the normal allele (O'Hoy et al. 1993). This occurred in a rare instance in which reduction was into the normal range, something which has been reported from paternal transmissions only (Shelbourne et al. 1992; O'Hoy et al. 1993; Brunner et al. 1993b). In other cases, the process which leads to expansion, such as unequal sister chromatid exchange or replication slippage, may be reversed or there may be direct deletion of repeats. The apparent instability threshold of 30–50 repeats correlates well with the energy required to form hairpin loop structures (Gacy et al. 1995). Jansen et al. (1994) suggested that longer repeat lengths allow the formation of elaborate mispaired structures, which may act as pause sites during replication and therefore need to be resolved before gametogenesis can proceed.

Based on the above, it has been suggested that the restriction of cDM to maternal transmissions is primarily due to the inability of sperm to support the necessary large expansions, and is not necessarily the result of some unique mechanism or mechanisms that allow for large expansions in female transmissions (Lavedan et al. 1993b; Mulley et al. 1993). However, other factors have been considered in the exclusive transmission of cDM through mothers, such as the lowered fertility of affected males and the paucity of small alleles in affected females. Moreover, exceptional cases of very large transmissions from males have been seen which did not result in cDM (e.g., as reported by Passos-Bueno et al. 1994). The possibility exists that a maternal uterine factor contributes to cDM. A greater risk of giving birth to a child with cDM was noted for mothers who had already borne a cDM-affected child (Harper 1989). A related familial risk for cDM has also been defined. In addition, a greater risk that an affected baby would have cDM was seen for mothers who manifested multisystemic disease prior to or during pregnancy (Koch et al. 1991). However, as referred to above, these outcomes may simply be related to the likelihood of longer repeats being present in the oocytes of mothers defined as being at risk. An intrauterine factor has been proposed to play a role in the generation of cDM fetuses, and some experimental evidence has been provided using newborn rats as a test system (Harper and Dyken 1972; Farkas-Bargeton et al. 1988). However, the mechanism of such a factor would have to involve an interplay between the CTG repeat status in the mother and infant and complete tolerance of the factor by all unaffected (normal CTG repeat) infants. Rare cases of paternally transmitted cDM have now been seen, which also places this in doubt (Bergoffen et al. 1994; Nakagawa et al. 1994; Ohya et al. 1994). The possibility that genomic imprinting affects the way CTG repeats expand after fertilization or affect the developing infant has also been considered, but no allelic sex differences in methylation patterns (Shaw et al. 1993a; Ashizawa et al. 1994a) or in mRNA expression patterns (Jansen et al. 1993) have been seen.

3
Gene Structure and Expression

3.1
Gene Structure

Identification of the unstable fragment at the DM locus was quickly followed by characterization of the transcribed region which contained the expansion. cDNA was cloned from human and mouse heart, muscle, and brain libraries (Brook et al. 1992; Fu et al. 1992; Jansen et al. 1992). For human DMPK, the transcripts were oriented in a 5'-telomeric to 3'-centromeric direction. Transcripts were also identified for a gene lying very close to the DMPK gene, just 1.1 kb upstream, termed DMR-N9 in the mouse or 59 in humans. Human and mouse genomic sequences were determined (Jansen et al. 1992; Fu et al. 1993; Mahadevan et al. 1993a; Shaw et al. 1993b). A complete open reading frame (ORF) was deduced from overlapping cDNA, with the translation initiation codon originally positioned by extension of the ORF upstream using genomic sequence homology between human and mouse (Jansen et al. 1992). Intron-exon boundaries were determined and 15 exons identified over an area of approximately 14 kb (Jansen et al. 1992; Mahadevan et al. 1993a; Shaw et al. 1993b). An ORF was identified encoding 629 amino acids, with a predicted weight of approximately 69 kDa (Jansen et al. 1992; Mahadevan et al. 1993a), which varied only slightly in other reports (Shaw et al. 1993b, Sasagawa et al. 1994). This structure was later confirmed by the isolation of full-length cDNA spanning the entire coding region (Sasagawa et al. 1994; Whiting et al. 1995). A consensus polyadenylation signal was present after 735 nucleotides of 3'-UTR. The longest cDNA identified in our laboratory included 326 nucleotides of 5'-UTR, giving a total transcript size of 2.9 kb or more, consistent with the unique species seen on Northern blot analysis with a size of 3.0–3.3 kb (Brook et al. 1992, Jansen et al. 1992).

Conceptual translation of the ORF reveals a protein with discrete domains (Fig. 6). The first exon is leucine rich and has a periodic sequence consistent with a leucine zipper motif, although no data has yet been provided to support this. Domain homology with various thymopoietins has also been suggested (Sasagawa et al. 1994). Based on a high overall hydrophobicity, a signal sequence function was proposed (van der Ven et al. 1993). Exons 2–8 have strong sequence homology with various kinases. Originally, the greatest similarity was to vertebrate cAMP-dependent protein kinases. All 12 conserved subdomains of kinases and the signature sequence of serine/threonine specificity can be identified. Within exons 9–12 weaker homology to filamentous proteins such as myosin heavy chain (MHC) can be detected. A strong tendency to form a coiled-coil structure is predicted between residues 460 and 530 (ISREC Coils server). Exons 13–15 have no known homologies, but exon 15 has a hydrophobic sequence predicted to act as a transmembrane anchor (method of Argos and Rao 1986).

Fig. 6. DM protein kinase (DMPK) Domain Structure. *Boxes* represent the exons of the DMPK gene. Exons shaded *black (2–8)* contain all 12 conserved subdomains of protein kinases and contain the sequence signature for Ser/Thr kinases. The exons shaded *gray* have weak homology to myofibrillar proteins such as myosin heavy chain, and residues within exons 11 and 12 are predicted with high certainty to form a coiled-coil domain. Exon 15 *(cross-hatched)* contains a hydrophobic domain predicted to form a transmembrane helix, although this function was not specified in a translocation assay system. (See text for details)

A recent search of the database produces a list of kinases with sequence homology, many of which are not known to be cAMP-dependent, originating from a variety of organisms (Fig. 7). At the time of writing, a human kinase, PK428, whose function has not been described, has the highest overall similarity to DMPK. It is followed by a family of rho-binding kinases (e.g., ROCK or ROK-α, and ROK-β), which are known to be activated by rho, a small GTP binding protein involved in transducing signals which ultimately result in reorganiza-

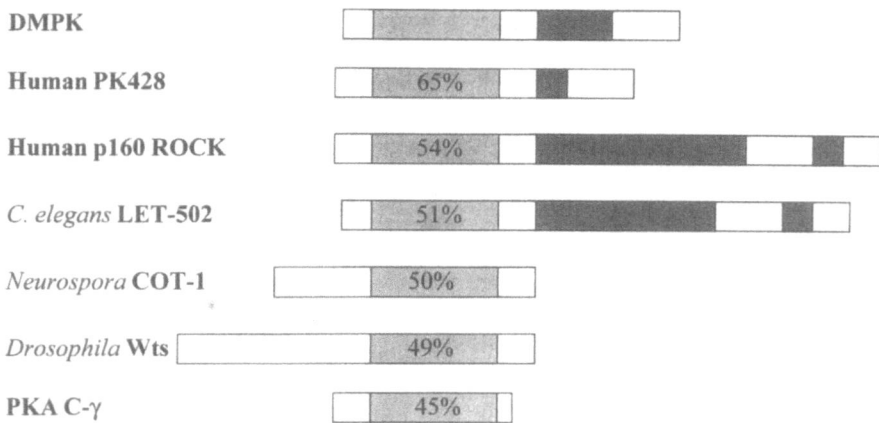

Fig. 7. Sequence alignment of DM protein kinase (DMPK) with related serine/threonine kinases, showing different kinases which have sequence similarity to DMPK. The percentage identity of the kinase domains *(light gray regions)* to DMPK is given. Human PK428 is a novel kinase with unknown function. DMPK and PK428 have additional sequence conservation within their respective first exons (N-terminal to the kinase domain). p160 ROCK (rho-binding kinase) and presumably *C. elegans* LET-502 are activated by rho. Both are believed to act as negative regulators of myosin phosphatase. They contain a C-terminal pleckstrin homology domain *(striped region)* not present in DMPK. All of the above contain a region predicted to form a coiled-coil *(dark gray boxes)* of varying size C-terminal to the kinase domain. Other more distantly related kinases originate from a diverse variety of organisms

tion of the cytoskeleton (see Lim et al. 1996). Although rodent and human rho kinases are most similar to DMPK, strong evolutionary conservation is evident in this family, as a presumed *Caenorhabditis elegans* rho kinase (LET-502), which is involved in the process of embryonic elongation (Wissman et al. 1997), also shows homology to DMPK. Moreover, while homology is primarily contained within the kinase domain, a common structural organization can be seen for DMPK, PK428, and the rho kinases in which an N-terminal kinase domain is followed by a coiled-coil domain. For DMPK and PK428, the predicted coil is modest (between 40–50 residues) while the rho kinases have much longer coils (up to 600 residues). Finally, PK428 alone has further homology with DMPK, beginning from their respective initiation codons and proceeding through the leucine-rich exon 1. Other kinases with homology include species from plants (common ice plant, common tobacco, spinach), Ndr nuclear protein kinases, from protozoa (*Euplotes crassus*), from yeast (YNL161W gene), the colonial growth mutant Cot-1 from *Neurospora crassa*, and the tumor suppressor *warts* from *Drosophilia*. DMPK can therefore be considered the founding member of a novel group of kinases.

3.2
Transcript Expression

Northern blot analysis of the human DMPK transcript (Jansen et al. 1992; Brook et al. 1992; Hoffman-Radvanyi et al. 1993) detected abundant message in striated muscle, heart, and organs containing smooth muscle (lung and bladder). Notably, this pattern closely parallels the tissues which are predominantly affected by disease. Low levels were also detected in kidney, liver, spleen, thymus, testis, seminal vesicle, and ovarium (Jansen et al. 1992) and in brain (Brook et al. 1992). No expression was detected in fetal stomach, small bowel, cerebellum, or liver or in adult liver (Hoffman-Radvanyi et al. 1993). Reverse transcriptase (RT)-PCR analysis detects transcripts in many more tissues, both adult and fetal (Jansen et al. 1992; Fu et al. 1993). DMPK expression appears to be widely conserved in vertebrates from studies of transcript or protein expression or from Southern blot detection of cross-species sequence conservation (Brook et al. 1992; Whiting et al. 1995; Dunne et al. 1996b). These observations, in conjunction with the widespread tissue distribution, suggest a ubiquitous function for DMPK. During development in the mouse, DMPK message can be detected by in situ hybridization at day 10.5 of embryogenesis in a restricted region of the somites (Jansen et al. 1996). Expression in all striated and smooth muscle was visible by day 14.5, and expression in all muscle groups increased and remained high throughout life. Interestingly, expression in neural tissue was not seen until after birth: at 14 days postnatally, transcript was detected in the retina, cerebellum, and hippocampus. This suggests a fundamentally different contribution of DMPK to these tissues. A study of human preimplantation embryos detected the onset of embryonic DMPK expression very early, at the one- to four-cell stage (Daniels et al. 1995).

Analysis of cloned cDNA also revealed a variety of alternatively spliced forms (Fig. 8; Jansen et al. 1992; Fu et al. 1993; Mahadevan et al. 1993a). These were more abundant in the mouse due to the use of cryptic splice sites. In some cases, intronic sequence was used to create additional coding sequence, in others cases, complete exons or portions thereof were removed. In humans, the alternate transcripts most commonly observed involved deletion of either exon 13 or 14, or both. Removal of small C-terminal portions of exon 8 from mouse and human transcripts was also seen in different studies (Jansen et al. 1992; Fu et al. 1993). By RT-PCR analysis, many different products using alternate consensus splice donors and acceptors were also detected at the 5' end (Fu et al. 1993). While some of these alternate products may be due to cDNA cloning artifacts or may represent amplification of incompletely spliced intronic sequence, the consistent detection of forms with exons 13 and/or 14 completely excised has attracted attention. Removal of exon 13 and 14, or 14 alone, shifts the reading frame such that translation terminates after two codons of exon 15,

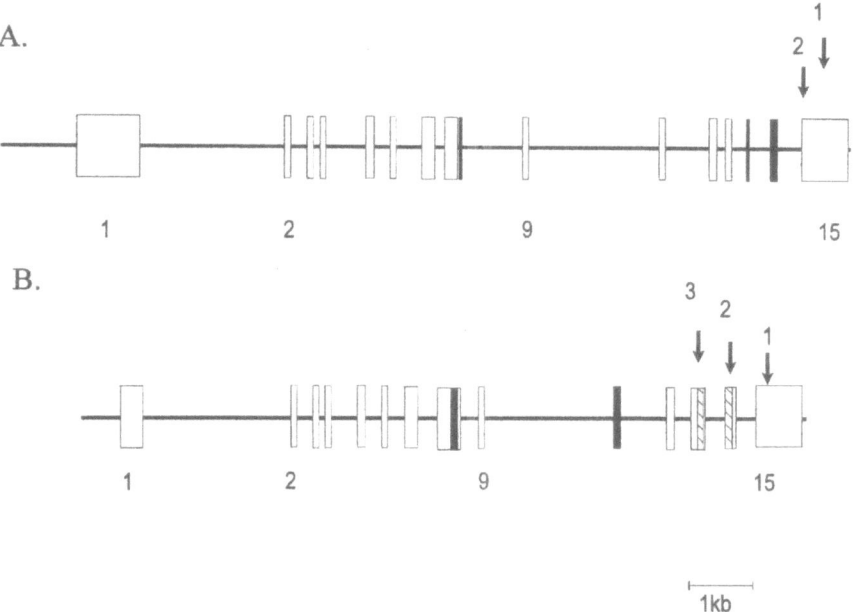

Fig. 8 A, B. Various splice variants for **A** the human and **B** the murine DM protein kinase (DMPK) genes. *Boxes* represent the exons of the DMPK genes. The *black boxes* represent exon regions deleted in various isoforms of the gene. The *hatched boxes* in the mouse gene represent intron regions used to generate truncated isoforms of the gene. The *numbered arrows* indicate the position of translational stop sites used by various isoforms. A Stop codon 1 is used by the full-length transcript. Stop codon 2 is used by a heart cDNA isoform in which exons 13 and 14 are deleted. The small portion of exon 8 variably present contains a potential glycosaminoglycan attachment site. **B** Stop codon 1 is used by the full-length transcript. Stop codon 2 is used by an isoform which includes a portion of intron 13. Stop codon 3 is used by an isoform which includes intron 12. (Mahadevan et al. 1993a)

thus truncating the product before the hydrophobic sequence therein. It has been proposed that this might create non membrane-bound forms of DMPK protein. The cDNA for these exon deletion forms have been found primarily in cardiac and brain libraries and in the corresponding tissues by RT-PCR (Gennarelli et al. 1995; Krahe et al. 1995b), perhaps suggesting a requirement for a more soluble protein. Removal of exon 13 shifts the reading frame, producing a basic, proline-rich sequence with homology to the Clox homeobox protein; this form was found to be specific to brain by RT-PCR (Genarelli et al. 1995). In addition, the C-terminal portion of exon 8 absent in some clones includes a potential glycosaminglycan attachment site and may therefore again produce protein isoforms which differ in localization or function. Direct verification of the expression of these proposed isoforms would greatly advance our understanding of DMPK function.

3.3
Protein Expression

The analysis of DMPK expression has been problematic and in part has led to some confusion regarding the effects of the CTG expansion on gene expression. However, the initial attempts to detect DMPK protein appeared quite consistent. Using an antipeptide serum, Fu et al. (1993) first reported the detection by immunoblot analysis of a 55 kDa protein in skeletal muscle, heart, and to a lesser degree in brain. This was followed by reports of species of 52 kDa (and 42 kDa; Brewster et al. 1993), 53 kDa (van der Ven et al. 1993), 54 kDa (Etongué-Mayer et al. 1994), 53 kDa (and 62 kDa; Koga et al. 1994), and 50–54 kDa (Salvatori et al. 1994), all using antisera to peptides. The similar sizes of the major species, and detection of these in the appropriate tissues, despite the use of immunizing peptides from many regions of the protein, were convincing in terms of the authenticity of a 50- to 55-kDa protein. This size range is smaller than predicted, although initiation within the first intron could produce such an isoform. Other possible explanations include anomalous mobility on SDS polyacrylamide gel electrophoresis (PAGE) or proteolytic processing. The latter would have to involve removal of approximately 14–17 kDa (likely too large for exon 1), presumably without ablating kinase function. In contrast, we detected two larger protein species using a polyclonal antiserum generated against a C-terminal fragment of the DMPK protein (αDMK). We estimated their sizes at approximately 74 and 82 kDa and found preferential expression of the larger species in skeletal muscle and of the smaller one in heart and brain (Whiting et al. 1995). These species were longer than predicted, possibly due to posttranslational modification. This issue of the correct identification of DMPK protein species is critical to ultimately determining the consequences of repeat expansion on gene expression. Reduced levels of their respective protein candidates were reported in myotonic patient muscle by some laboratories (Fu et al. 1993; Koga et al. 1994).

Recent studies have demonstrated that some of the 50- to 55-kDa candidates may not be DMPK isoforms, but cross-reactive proteins. Timchenko et al. (1995) found that a second antipeptide sera recognized a 72-kDa protein in addition to the 55-kDa protein, while an antifusion protein sera preferentially recognized the 72-kDa form. The 72-kDa species was shown to have kinase activity, while the 55-kDa one did not. Jansen et al. (1996) found that their 53-kDa protein did not display the appropriate pattern of expression in knockout and transgenic mice. Etongué-Mayer et al. (1994) found a tyrosine kinase activity for their 55 kDa protein isolated from skeletal muscle, despite the serine/threonine-specificity signature sequence present in DMPK and its demonstrated serine/threonine specificity (see below). Independent evidence of the authenticity of these candidates is therefore now required in light of this data. αDMK recognized a species overexpressed in transgenic tissues relative to wild type mice and absent in knockout mice (Jansen et al. 1996; Reddy et al. 1996). In addition, the larger human isoform (predominant in skeletal muscle) was found to comigrate with protein produced from full-length cDNA in cell culture (Maeda et al. 1995; Waring et al. 1996). The relationship of the two (74- and 82-kDa) isoforms is uncertain. The smaller (predominant in heart) comigrates with a species produced from an exon 13/14 deletion cDNA and may therefore be a product of differential splicing (Waring et al. 1996). Alternatively, the larger species may be a post-translational modification of the smaller. Protein sequencing will be required to resolve this. Epstein and colleagues have also reported the detection of larger isoforms – a 64 kDa species in muscle and a 79 kDa species in brain – using a monoclonal antibody (mAb DM-1) generated against full-length protein (Dunne et al. 1996a). Although these proteins did not match the size of control protein species expressed in COS cells, proteolytic digestion of the 64 kDa protein generated a similar pattern of peptides as a bacterially expressed control and therefore may represent another DMPK isoform. A 67 kDa protein was also reported in lens with a polyclonal antibody against full-length protein (DM-2; Dunne et al. 1996b).

Despite the distinct results obtained for antipeptide and antifusion protein sera in immunoblotting experiments, a surprising consensus was obtained for immunoflurescence patterns. Prominent staining was first noted by van der Ven et al. (1993) in neuromuscular (NMJ) and myotendinous junctions in rat, mouse, and human. A weak sarcolemmal reactivity was restricted to type I fibers. Staining was also seen in intercalated discs in cardiac muscle, dense plaques in smooth muscle, and in most neurons in the brain. Whiting et al. (1995) likewise found staining at NMJ and intercalated discs in cardiac muscle, a pattern verified by Maeda et al. (1995). Restriction of DMPK to the postsynaptic side of NMJ was further demonstrated. In the brain, prominent staining was observed in ventricular ependymal and choroid plexus cells as well as discrete neural populations in the hippocampus, cerebellum and medulla. Salvatori et al. (1994) again reported staining of NMJ, intercalated discs, and plasma membrane of skeletal muscle. Dunne et al. (1996a) reported mAb DM-1 reactivity in the sarcoplasm of skeletal muscle fibers, restricted to type I fibers. This was further localized to the I band, consistent with localization in the triad region. Using the anti-55 kDa antibody of Caskey and colleagues, Tachi et al. (1995) reported staining of NMJ, muscle spindle, and prominent sarcoplasmic staining of both type I and type II adult fibers. Pronounced staining of regenerating fibers present in biopsies of Duchenne/Beckers muscular dystrophy was also noted. The similarities in profiles for the antipeptide and antifusion protein sera may seem unlikely upon first consideration, but specificity can vary for antisera in different tests. Immunoblotting is prone to detection of cross-reactive species due to mass effects, because bound antibodies are removed from equilibrium. We suggest that the common sites recognized in the various immunofluorescence studies are accurate (NMJ and other peripheral aspects of skeletal muscle fibers, cardiac intercalated discs, neural populations) and that this pattern suggests the presence of DMPK at specialized sites of intercellular contact.

Biochemical fractionation of different cells and tissues expressing DMPK would help confirm its localization, but little data is currently available. Maeda et al. (1996) found that both DMPK isoforms were predominantly soluble, and the majority could be eluted from a preparation of intercalated discs, suggesting only peripheral association. We have also found that the majority of DMPK can be extracted by nonionic detergent from tissue samples, consistent with a cytosolic distribution (Whiting et al. 1995). However, when expressed in insect cells, full-length DMPK had a preferential association with membranous fractions not seen for the exon 13/14 deletion truncated form (Waring et al. 1996). The C-terminal hydrophobic region may therefore confer some association with a component of peripheral membranes. Solubility in alkaline carbonate buffer confirmed that DMPK is not an integral membrane protein. For their candidate (54 kDa), Salvatori et al. (1994) also reported a peripheral association with plasma membranes and with sarcoplasmic reticulum of predominantly fast skeletal and cardiac muscle (but not in triads). Other laboratories have reported predominantly soluble behavior of their candidate isoforms.

3.4
Protein Structure and Function

Little is known about DMPK secondary and tertiary interactions, post-translational modifications, and substrate specificity. Delineation of the normal role for DMPK would not only be interesting due to its unusual domain organization and ubiquitous expression, but would also allow us to understand the consequences of any perturbation of normal expression levels. No specific function can currently be deduced from the list of kinases with sequence similarity but, in general, the information available suggests some control over cell shape and/or morphogenesis. The rho-binding kinases can induce the formation of stress fibers (cytoskeletal filaments) in fibroblasts when expressed in an unregulated fashion (Leung et al. 1996). They are bound to and activated by GTP-bound (activated) rho. This has been shown to stimulate the kinase activity of ROK, which can phosphorylate the myosin-binding subunit (MBS) of myosin phosphatase thereby downregulating its ability to dephosphorylate myosin light chain (MLC). This series of events would ultimately result in greater phosphorylation of MLC, increasing contractibility of smooth muscle, or allowing the formation of stress fibers in non-muscle cells. Let-502 has a similar substrate specificity and likely a very similar function (Wissman et al. 1997). However, while a similar organization for DMPK and the rho kinases is evident, the latter have a C-terminal plekstrin homology domain containing a cysteine/histidine-rich sequence in addition to the longer coiled-coil. This may imply a distinct subcellular localization and possibly a divergent regulatory role. A rho-binding motif has been suggested for DMPK, but the sequence conservation is weak (Wissman et al. 1997). It will be important to determine whether DMPK and other members of this family are also activated by rho and regulate interactions of cytoskeletal or myofibrillar proteins. The Cot-1 gene product was identified in a temperature-sensitive mutant strain of the fungus *Neurospora crassa*, which had restricted colonial growth due to abnormal hyphal tip elongation (Yarden et al. 1992). The *Drosophilia warts* product is also required for normal cellular proliferation and morphogenesis (Justice et al. 1995). Homozygous loss of this gene results in overproliferation and hypertrophy at the apical surface of imaginal disc epithelial cells, causing wart-like growths on the adult. Based on this sequence similarity and the occasional presence of pilomatrixomas and other tumors in DM patients, a tumor suppressor function has also been suggested for DMPK (Justice et al. 1995; Harris et al. 1996).

DMPK expressed in bacteria or cell culture generally migrates slower than expected from its primary sequence, suggesting post-translational modification. Phosphorylation of DMPK expressed in cultured eukaryotic cells has been demonstrated (Timchenko et al. 1995; Waring et al. 1996) and has been inferred to be occurring for bacterially expressed DMPK based on autophosphorylation patterns (Dunne et al. 1996a). Another possible explanation might be glycosaminoglycan modification of exon 8-encoded sequence. In vitro kinase activity was also demonstrated in these three studies. Both autophosphorylation and transphosphorylation of foreign substrates were demonstrated, with a serine/threonine specificity (Dunne et al. 1994, Timchenko et al. 1995). Activity was not greatly inhibited by the compounds H7 and H8, which have specificity against cAMP-dependent protein kinase, cGMP-dependent protein kinase, and protein kinase C, or by inhibitors of casein kinases and calcium/calmodulin protein kinase. This again suggests that DMPK has properties that are distinct from those of these other kinase classes. We noted that removal of phosphates added in cell culture prior to an in vitro kinase assay decreased the activity observed, suggesting that DMPK is upregulated by phosphate groups. The residue or residues modified by phosphate and the structural consequences remain to be determined.

No change in the mobility of DMPK or protection from exogenous protease was found using in vitro translation reactions supplemented with canine pancreatic microsomes (Waring et al. 1996). This demonstrates that DMPK is not translocated, which is surprising in light of its structure, but consistent with its carbonate solubility. This also rules out glycosylation of DMPK as the basis for slow mobility. Dunne et al. (1994) noted that truncated DMPK produced in bacteria sedimented at approximately 41 S on velocity gradients, roughly equivalent to a composition of 21 subunits, given a spherical particle. In accordance, they observed regular spherical particles 30–50 nm in diameter by electron microscopy. We noted that DMPK produced both in insect cells or in COS cells migrated as large multimers on nonreducing SDS-PAGE and that the C terminus contributed to a higher-order structure (Waring et al. 1996). This might somehow be related to the distinct localization noted (see above) for full-length and truncated DMPK forms. The nature or purpose of these complexes is currently unknown, but might reflect assembly in vivo with other members of the myofibrillar or cytoskeletal apparatus. It is interesting to note that the small amount of DMPK in the heart which remained associated with an intercalated disc-enriched pellet consisted of the larger isoform (Maeda et al. 1995).

The nondystrophic myotonias and periodic paralyses are a group of disorders caused by impaired electrical excitability of the sarcolemma due to ion channel defects. Mutations have been identified in voltage-gated Na^+ or Ca^{2+} channels in different forms of periodic paralysis, and in the skeletal muscle chloride channel in myotonia congenita. A similar defect underlying myotonia has long been speculated for DM, which would presumably involve aberrant phosphorylation at a regulatory site of an ion channel. One intriguing study found that coexpression of DMPK with rat skeletal muscle sodium channels in *Xenopus* oocytes reduced the amplitude of Na^+ currents and accelerated current decay (Mounsey et al. 1995). Downregulation of DMPK might therefore explain heightened muscle excitability. A candidate approach has been initiated using in vitro kinase assays to phosphorylate potential substrates. Dunne et al. (1994) tested *Torpedo* acetylcholine receptor subunits γ and δ and the β-subunit of erythrocyte spectrin and did not observe any specific phosphorylation. Bush et al. (1996) reported that myogenin could serve as an in vitro substrate. Timchenko et al. (1995) found that recombinant DMPK could phosphorylate the β-subunit of the voltage-gated L-type Ca^{2+} channel (dihydropyridine receptor, DHPR) and that DMPK could be coprecipitated with DHPR from human cells. These two different candidates obviously imply quite different roles for DMPK; therefore, confirmation of these as DMPK substrates in vivo now becomes an essential step.

The underlying basis for the histological changes seen in adult DM muscle is not known, but may reflect the presence of regenerating fibers. The appearance of cDM patient muscle is more hypotonic than myopathic, and the abnormal presence of myotubes, satellite cells, small fiber size, and the incomplete differentiation of fiber type indicate a marked delay in muscle differentiation (Harper 1989). The expression of DMPK in muscle cell precursors has therefore been examined in an attempt to place DMPK in the context of muscle differentiation. P19 stem cells, a pluripotential embryonic carcinoma line, were found to express DMPK de novo when expressing the myogenic regulator MyoD in a stable fashion (Skerjanc et al. 1994). Aggregation of these cells, an important cofactor in skeletal muscle differentiation in this system, did not further enhance the expression of DMPK or *myf*-5 (another myogenic regulator). These genes can therefore be considered markers of myoblasts, but not muscle fibers. In addition, fibroblasts from normal and DM patients transformed with MyoD showed greatly increased levels of DMPK transcript over wild-type cells by Northern blot analysis (Otten and Tapscott 1995). The DMPK gene has E-box elements within its first intron, an observation made for other genes expressed in muscle. In in vitro studies, we have found that reporter genes under the control of the DMPK 5'-UTR, in combination with the first intron, have greater activity in muscle compared to non-muscle cell lines and are responsive to MyoD (our unpubl. observ.). DMPK is therefore directly or indirectly regulated by MyoD, consistent with, but not sufficient to explain the upregulation of DMPK in various tissues seen in the developing mouse embryo (Jansen et al. 1996). Placing a reporter gene under the control of a fragment from the 5'-UTR resulted in expression restricted to neural tissue in transgenic mice (King et al. 1996), consistent with the notion that the first intron is key to muscle-specific expression.

In a more direct study of DMPK function, undifferentiated (fibroblast-like) BC_3H1 cells which expressed DMPK in a stable fashion exhibited a myoblast-like cell morphology, similar to that of BC_3H1 cells induced to differentiate by serum withdrawal (Bush et al. 1996). Gene expression profiles for DMPK-transfected and differentiating cells were found to be similar by mRNA differential display. Dramatic increases in expression of skeletal muscle β-tropomyosin, myogenin, and creatine kinase M and decreases in expression of vimentin and retinoblastoma were seen in these lines, consistent with the activation of a skeletal muscle program. These studies demonstrate upregulation of DMPK early in myogenesis and responsiveness to myogenic regulators, possibly in keeping with some role in initiating or maintaining skeletal muscle identity. It will be interesting to determine how DMPK expression is regulated in other tissues, especially in light of the temporal difference seen between mouse muscle and neural tissue, and what the common functional requirement for DMPK might be.

4
Consequences of Repeat Expansion

The consequences of CTG repeat expansion with respect to the expression of DMPK, and indeed other genes in the vicinity, constitute a key area of research for DM. There are several possible ways in which the trinucleotide expansion might affect normal gene expression, including the following: (a) inhibition of DMPK transcriptional initiation or rate, (b) changes in transcript stability, processing, or export, (c) changes in translational efficiency, (d) interference with the expression of a neighboring gene, or (e) longer-range effects mediated by alteration in chromosome structure. Two opposing major models have been proposed which would explain the dominant mode of inheritance, i.e., down-regulation of DMPK expression (haploinsufficiency model) or a gain of function due to some novel property conferred on DMPK transcripts by the expansion. Both models have substantial uncertainties. The haploinsufficiency model has been questioned on the grounds that it would be unreasonable that variation of from 50 to 100 % of the normal expression levels could account for the wide range of clinical severity (Fischbeck 1994). This implies exquisite sensitivity to the amount of DMPK present. Another critical observation is that no other coding mutations which directly knock out the kinase function have been identified as a basis for DM. A model based on a transcript gain of function, presumably a toxic effect acting in *cis* and/or in *trans*, would appear to be novel in human disease. It is uncertain how a generalized dysfunction of cellular metabolism of this sort might account for the wide range of pathologies noted in DM. There is a precedent for the regulation of normal cellular metabolism by transcripts, as muscle-specific genes such as tropomyosin inhibit growth and promote differentiation via their 3' UTR (Rastinejad and Blau 1993). Brook et al. (1992) have also pointed out gain-of-function mutations present in the 3' UTR of some *C. elegans* genes. It will be crucial to determine which one (or both, in combination) of these two possible mechanisms is the basis for DM before treatment strategies can be devised.

4.1
Effects on Transcript Levels and Metabolism

Early studies which analyzed DMPK transcript levels in muscle by RT-PCR found substantially decreased levels of mutant transcripts, supporting the haploinsufficiency model. Fu et al. (1993) used cDNA primed with both oligo-dT and random primers and found that total DMPK transcript levels were lowered in adult patient samples, although a wide range was seen in control samples. Specific primers were used which flanked the CTG repeat region in order to distinguish the alleles, and the reduction was found to be due to the mutant allele, with the levels detected showing a negative correlation with the expansion size (see also Novelli et al. 1993). More dramatic results were found when somatic cell hybrids containing chromosome 19 from an adult DM patient (113 repeats) or a normal control (13 repeats) were analyzed (Carango et al. 1993). This approach allows the level of mutant transcripts to be determined directly and in the absence of any effects in *trans* of one allele on the other, which has been proposed (see below). Regions upstream of the CTG expansion, specific to total or spliced DMPK transcripts, were amplified using the appropriate primers from either oligo-dT or random-primed cDNA. A reduction in unprocessed mutant transcript to approximately 18 % of the control levels was seen, and no processed transcript could be detected. In another study, northern blot analysis of RNA prepared from a 20 week-old fetus (1.8-kb expansion) or from two cDM infants (6.4- and 8.0-kb expansions, respectively), also failed to detect expanded transcripts (Hoffman-Radvanyi et al. 1993). In addition, the normal allele appeared reduced to 22 % of that of a normal fetus, instead of the expected 50 %, implying that the mutant allele can interfere with the normal allele in *trans*. In agreement, RT-PCR analysis of the fetal sample using random-primed cDNA amplified in three different regions (upstream, downstream, and across the repeats) again showed reduction of total DMPK transcript to approximately 30–40 % of normal.

These studies were therefore consistent in their detection of mutant transcripts at low levels. In contrast, Sabourin et al. (1993) reported strikingly different finding for transcript levels from various cDM tissues. Total transcript levels, as analyzed by competitive RT-PCR and slot-blot analysis, were increased over controls, estimated at two- to fourfold for heart and brain, four- to sixfold for liver and lung, and 14-fold in brain, although the appropriate age-matched controls were not available. Wild-type and mutant alleles were distinguished using an informative biallelic polymorphism (Mahadevan et al. 1993a), and transcripts from both alleles were found to be expressed at equal levels. One other recent publication using an allele-specific RT-PCR system also found no difference in mature transcript levels for the mutant and normal alleles in adult muscle samples, regardless of repeat size (Bhagwati et al. 1996a). In addition, large mutant transcripts were detected by Northern blot in cDM patient fibroblasts (10 kb in size; Sabourin et al. 1993) or in fibroblasts from a DM patient after conversion to muscle by MyoD (6 kb; Otten and Tapscott 1995). The basis for these discrepancies is uncertain. Roses (1994) and Sabourin et al. (1993) cautioned against measurements made for mutant transcripts based on RT-PCR which proceeds through the CTG repeat region. Inefficient polymerization through templates containing trinucleotide repeats, especially for RT, has been a very common finding. This would not affect the measurement of total transcript levels in the study by Fu et al. (1993), but would affect their direct measurements of mutant allele levels. Carango et al. (1993) amplified regions very close to the CTG repeats (within ten nucleotides). Therefore, a restriction of potential RT primers to those which must proceed completely or in part through the CTG repeats could explain the very low transcript levels determined (for a relatively small repeat expansion). However, it is also possible that very little transcript is produced by hemizygous hybrid cells (see below). Sabourin et al. (1993) used DMPK-specific RT primers upstream from the repeats. However, there were other important differences between Sabourin et al. (1993) and the previous two studies, as lymphoblastoid lines (as opposed to muscle) were used in allele-specific RT-PCR experiments, and cDM (as opposed to adult) tissues were used throughout.

Two more recent, detailed studies, using allele-specific quantitative RT-PCR systems very similar to that of Sabourin et al. (1993), carefully avoided these difficulties with the CTG repeat region to measure transcripts from both the precursor and mature fractions from a variety of sources. Wang et al. (1995) analyzed both total and poly(A)$^+$ RNA in muscle from adult patients using DMPK-specific RT primers. In total RNA from patients' muscle, a reduction to approximately 50 % of control levels was seen for DMPK transcripts, with considerable overlap between the two distributions, similar to the results obtained by Fu et al. (1993). However, a similar decline was seen in myopathic control samples. In contrast, there was a striking paucity of DMPK message in the poly(A)$^+$ fraction, with mutant message decreased to approximately 25 % of the normal allele levels. The absolute levels of transcript indicated that 86 % of the mutant transcript and 70 % of the normal transcript were lost in mature mRNA (as opposed to about 10 % in unaffected controls).

Krahe et al. (1995b) used a system specific for either spliced or unspliced message (with either DMPK-specific primers or a combination of oligo-dT and random primers for cDNA synthesis) to analyze muscle samples, cell lines, and somatic cell hybrids derived from both adult and cDM patients. They observed no significant deviation from a ratio of 1 for unprocessed mutant and normal allele transcripts from any source. Again, however, a reduction in mature mutant transcripts relative to normal transcripts (mean ratio, 0.49) was observed, which could be correlated with the expansion size. Unlike the results obtained by Wang et al. (1995), however, absolute levels of mature mRNA did not decline, except for a homozygous patient and in somatic cell hybrids. This suggests instead that total locus activity only declines in the absence of the wild-type allele, consistent with the results obtained by Carango et al. (1993), implying an interaction between alleles. Therefore, while these two studies differ in their conclusions regarding total and mature DMPK transcript levels, the total weight of the results strongly suggests that CTG expansions do not affect transcription, but rather some aspect of transcript metabolism. Krahe et al. (1995b) were not able to see any difference in stability or splicing patterns between mutant and normal transcripts, suggesting instead that export may be involved.

The validation of an effect of the CTG expansion on posttranscriptional processing would be a critical clue to the pathogenesis of DM. One intriguing study examined the distribution of DMPK transcripts in fibroblasts and muscle biopsies from adult patients by in situ hybridization. Mutant transcripts were found aggregated in discrete foci in the nuclei of affected cells (Taneja et al. 1995). The nature of these foci is not currently understood. They may be indicative of an irreversible association with nuclear factors or structures or of structural interference with export due to aberrant folding specified by the repeats. In any case, a physical basis for the paucity of mutant message in the mature cytoplasmic fraction now seems apparent, and mechanisms can be envisioned by which the mutant DMPK message might affect transcript export both in *cis* and in *trans* (Wang et al. 1995; Krahe et al. 1995b). Critically, this mechanism is still consistent with the fact that affected tissues have the highest levels of DMPK expression. It could also be speculated that difficulties in extracting the aggregated mutant message might contribute to the difficulty in detection by Northern blot analysis (Hoffman-Radvanyi et al. 1993).

An attempt to colocalize these foci with known splicing factors was not successful. However, gel shift assays designed to investigate factors which can bind various triplet repeats in vitro led to the identification of CUG binding proteins involved in two RNA-protein complexes (Timchenko et al. 1996a,b). These activities (CUG-BP1, CUG-BP2) were distributed between the nucleus and cytoplasm, depending upon the cell type. The heterogeneous nuclear ribonucleoprotein (hnRNP) hNab50 was detected in both of these complexes and appears to consist of two isoforms responsible for CUG-BP1 and CUG-BP2. Furthermore, binding of hNab50 to the 3' UTR of DMPK was demonstrated in vivo, and the CUG-BP activities was suggested to be altered in DM lymphocytes. The authors suggested that CTG expansion might greatly increase the available binding sites for hNab50, a member of a class of proteins thought to be essential for pre-mRNA processing, effectively titrating it in the various affected tissues. Another study has identified two proteins (approximately 25 and 35 kDa) which bind CUG sequences and were also speculated to be involved in pre-mRNA processing (Bhagwati et al. 1996b).

4.2
Effects on Protein Expression

Obviously, determination of the level of DMPK protein in affected tissues is a direct way to confirm or refute disease models. However, clear conclusions cannot yet be drawn from the studies to date due to the difficulties in identifying DMPK isoforms. Candidates in the smaller size range (50–55 kDa) were reported to be decreased in patients' muscle samples (Fu et al. 1993; Koga et al. 1994). Subsequent analysis of the 72-kDa species recognized by the antisera of Caskey and coworkers found no decrease in expression in patient muscle, regardless of repeat size (Bhagwati et al. 1996a). Using αDMK antisera from our laboratory, a decrease in both isoforms relative to total myosin was observed in skeletal muscle of two adult patients and relative to cadherin in the heart of another (Maeda et al. 1995). Our own results suggest that levels are decreased in skeletal muscle relative to actin, but not relative to MHC I, a marker for the type I fibers which are believed to express DMPK and which undergo selective atrophy in diseased muscle (unpubl. observ.). Likewise, Epstein and coworkers (Dunne et al. 1996a) found no significant decrease in DMPK (64 kDa species recognized by mAb DM-1) in patient muscle. Instead, levels were increased relative to MHC I in severely affected distal muscle, where type I fiber atrophy was pronounced. Variations in tissue composition with aging and from the progressive wasting often present in DM are therefore important variables which must be considered. The distribution and intensity of DMPK staining by immunofluorescence, which may be more reliable at this point, were not different in patient tissues in one study (van der Ven et al. 1993). However, staining with mAb-1 was found to redistribute from triadic regions to sarcoplasmic masses (which are characteristic of DM) in affected distal muscle (Dunne et al. 1996a). Redistribution of DM-2 reactivity (67 kDa) in the eye from a perinuclear region to the nucleus of lens epithelial cells has also been reported (Dunne et al. 1996b). In summary, evidence suggests that modest declines in DMPK levels and redistribution in discrete areas may occur in adult patients, but perhaps only secondary to degenerative processes. Due to the recent progress in the validation of antisera, studies should yield key results in this area in the near future; cDM samples will be particularly informative.

4.3
Neighboring Gene Expression and Chromatin Structure

The inability to clearly associate the expansion of the CTG repeat region with changes in DMPK expression has led some groups to look farther afield. Large expansions might dramatically alter chromatin structure, affecting gene expression at a distance, or perhaps disrupting replication timing as demonstrated for the FMR-1 gene (Hansen et al. 1993). The DMPK gene is located within a gene-rich area of chromosome 19. The upstream DMR-N9/H59 gene is extremely close to the DMPK gene, which has led to speculation that they may be coordinately regulated (Jansen et al. 1992; Shaw et al. 1993b). The function of this gene is unknown, but four regions contain some homology to WD repeats, which are found in a diverse family of proteins involved in activities such as G protein signal transduction, cell division, and transcription (Jansen et al. 1995). The highest expression levels are in brain and testis, so alterations in N9/H59 expression could potentially contribute to certain aspects of the DM phenotype in more severe cases. However, in preliminary studies, no decreases in H59 expression were seen in DM patients (Funanage et al. 1996) or in cytoplasmic transcript levels in patient fibroblasts (Hamshere et al. 1996).

The CTG repeat region lies within a large CpG island which extends further 3' into an area with cross-species sequence conservation, indicating downstream transcribed sequences. Examination of genomic sequence revealed three potential exons with high homology to mouse sequence, very close to the DMPK gene (Boucher et al. 1995). RT-PCR detected the appropriate products from this region in heart, skeletal muscle, and brain. However, cDNA clones could not be identified in screens of several cDNA libraries, and expression was reported to be extremely low by Northern blot analysis. The first exon contains a region homologous to a homeodomain found in genes such as *Drosophilia sine oculis* and the mouse AREC3 and *six* genes. This new ORF was therefore termed the DM locus-associated homeodomain protein (DMAHP). Interestingly, AREC3 is expressed in C2C12 mouse myoblasts and upregulated during differentiation of these cells (Kawakami et al. 1996). It has been proposed that CTG repeat expansion may disrupt the promoter of DMAHP, thereby acting as a transcriptional silencer. This might interfere with embryonic development, although the effects of this would obviously become apparent only much later in adult-onset DM, or in later differentiation of adult tissues. Again, preliminary RT-PCR studies were inconclusive, with one suggesting a decrease of DMAHP transcripts in affected tissues (Thornton et al. 1996) and another reporting no change (Hamshere et al. 1996). Therefore, while the proximity of these genes to the repeat expansion and their expression patterns are intriguing, further evidence is required before they can be included in a mechanism of disease for DM.

A DNAse I-hypersensitive site has been detected within the CpG island, just downstream of the repeat region, perhaps within the regulatory region of DMAHP (Otten and Tapscott 1995). In patient cells, an allele containing an approximately 6 kb expansion was found to be relatively resistant to digestion, indicating a change in chromatin structure. Activation of DMPK transcription in fibroblasts by MyoD greatly increased the sensitivity of the wild-type allele, but the expanded allele remained resistant. Thus repeat expansion appears to dictate a more compact chromatin structure without silencing the DMPK locus. It was speculated that, apart from proposed effects on the downstream DMAHP gene, the induction of heterochromatin could affect gene expression over a large region. Variable activation or repression of different genes in the region might then contribute to the wide variation in the phenotype. Repetitive elements have been shown to position nucleosomes to maintain an open chromatin structure at an adjacent site in the absence of transcription. Repeat expansion might therefore occlude this region by excess nucleosome positioning.

Using an in vitro reconstitution system, it was demonstrated that assembly of nucleosomes was enhanced at CTG repeats and that this effect became more pronounced as the repeat size increased (Y-H Wang et al. 1994). In fact, 130 CTG repeats were nine times stronger than the *Xenopus borealis* 5 S RNA gene, previously the strongest natural nucleosome positioning element (Wang and Griffith 1995). It was also suggested that the threshold at which TNR disorders become apparent (40–50) might coincide with the number of base pairs required to form a nucleosome (146), implying a role for local chromatin structure in instability as well as transcriptional modulation. These reports, while circumstantial, do suggest an influence of the DM repeat on local chromatin structure in vivo. However, again, all changes secondary to chromatin structure must be viewed in the light of DM as a dominant disease. With current results suggesting little dramatic change in DMPK protein levels, it must then be speculated that many distal genes vary in their total expression levels (presumably 50–100 % of normal).

5
Models

5.1
Models of Disease

An in vivo model of DM would be invaluable in unraveling the apparently complex disease etiology. As discussed, the fact that multiple tissues are affected by a trinucleotide expansion at a single locus suggests either the loss of a relatively ubiquitous function, that multiple functions are modulated at a distance, or that a gain-of-function mutation can poison an essential metabolic activity in all tissues where DMPK is expressed. An animal model would allow investigators to identify common and disparate features in affected tissues, ultimately allowing conclusive support for one of the above possibilities. Recently, a series of studies has reported on attempts to model DM in the mouse by altering DMPK expression, the obvious first choice for study. Both overexpression of the human DMPK gene in the mouse and "knockout" of the endogenous mouse DM gene have been reported. The value of examining the contexts of both overexpression and underexpression at this time is clear, as neither an alteration in DMPK transcript steady state levels or *trans*-acting toxic effects of improperly processed transcripts can yet be conclusively supported.

Ablation of mouse DMPK expression by genetic knockout resulted in no apparent phenotypic abnormalities in one study (Jansen et al. 1996). MHC composition, fiber-type distribution and size, and electromyography profiles of knockout mouse muscle were not different from those of controls. No involvement of other systems or changes of a progressive nature were observed. The authors did note that calcium (Ca^{2+}) homeostasis in cultured myotubes derived from knockout mice was altered. It has been shown that myotubes from DM patients have higher intracellular $[Ca^{2+}]$ in the presence of extracellular Ca^{2+} (Jacobs et al. 1991). This was due to a difference in an activity sensitive to nifedipine, a specific inhibitor of L-type voltage-gated Ca^{2+} channels.

In a contrasting study, Reddy et al. (1996) demonstrated a muscle weakness in nullizygous mice that became apparent as they aged and appeared to be distal to Ca^{2+} release. Variations in fiber size, a modest increase in MyoD levels, and fibers positive for embryonic myosin were detected in older mice, indicative of muscle fiber regeneration (no change in MyoD levels were detected by Jansen et al. 1996). At the ultrastructural level, a loss of fiber structural integrity was observed, including distortion of the Z line and loss of myofibrillar organization. Again, no other systemic involvement was observed. No basis is apparent for the differences in these two mice lines, apart from strain differences, as both were thoroughly examined at the histological level. Of the changes noted, only increased fibrosis and variations in fiber size are (less prominent) features of DM (Harper 1989). Therefore, while these mice may eventually allow greater understanding of the normal structural/metabolic role of DMPK, it cannot be concluded at this time that these animals are models for DM (Hamshere and Brook 1996; Harris et al. 1996). However, as the authors of both studies point out, the *mdx* mouse presents a much milder phenotype than seen for human Duchenne muscular dystrophy patients, and interpretations must be made with caution.

Overexpression was studied using a 14 kb human genomic fragment, including all 15 DMPK exons and the last exon of the upstream DMR-N9/59 gene (Jansen et al. 1996). This fragment contained all *cis*-acting tissue-specific regulatory elements, as assessed by the pattern of detectable human DMPK transcripts in mouse tissues. In addition, overexpression of human protein five- to tenfold the level of endogenous mouse protein was seen in several tissues of a transgenic line (Tg26) bearing the highest copy number (approximately 25). However, no differences with control mice were noted for several measures of muscle function, including action, resting, and endplate potentials and electromyographic profiles. The appearance and MHC composition of striated and smooth muscle was normal, and NMJ exhibited no abnormalities. The authors did note an increased neonatal mortality and a hypertrophic cardiac myopathy (not similar to the fatty infiltration and fibrosis of myocardium and conductive tissue in DM) that were correlated to transgene copy number.

In surprising contrast, lines of transgenic mice recently developed in our laboratory using an identical genomic fragment showed several distinctive features of DM in skeletal muscle (our unpubl. observ.). Most conspicuously, a statistically significant atrophy of type I fibers was detected in skeletal muscle of transgenic mice when compared with control littermates. In addition, a significant increase in the centronucleation of muscle fibers and the presence of abnormal sarcoplasmic masses in some fibers were noted. While none of these changes are in themselves specific to DM, together they form a profile commonly seen in affected muscle and considered diagnostic for DM (Harper et al. 1989). We have also established myoblast lines from our transgenic mice and found a significant inhibition of myotube formation when compared to control mice. Importantly, like the changes noted above for the other transgenic model, the incidence of all these features was positively correlated with transgene copy number in different lines.

Matrix attachment region sequences (MARS) were used in the construction of our mice, which promote integration site-independent increases in transgene expression (Stief et al. 1989). We noted a striking transcript overexpression (25-fold) for our strain A mice (60 transgene copies), while relative protein levels (sixfold) were comparable to those of Tg26 for Jansen et al. (1996; five- to tenfold). Although transcript levels were not presented in the paper by Jansen et al. (1996), we speculate that the MARS sequences resulted in higher absolute transcript levels and that this accounts for the differences between these two mouse models. If so, this data greatly favors proposed *trans*-dominant toxic effects of overexpression of DMPK transcripts. However, this implies that excessive transcript with a normal repeat allele size is mechanistically equivalent to reduced or more nominal levels of transcripts with expanded repeats. As pointed out (Jansen et al. 1996), the extreme levels of DMPK products in these different mice are not comparable with the subtle changes observed in humans manifesting disease. If titration of a factor with CUG-binding specificity such as hNab50 (which has an equivalent activity in the mouse) is central to disease, this would reconcile our data with the requirement for trinucleotide expansion in human disease. However, further study, especially electromyographic profiles, will be required before these mice can be accepted as a DM model.

Surprisingly, a recent publication describes an animal model which almost certainly has a similar physiological basis as human DM. The mutant LWC strain of Japanese quail (Braga III et al. 1995) has difficulty lifting its wings ("locked wing cross"), which improves after repetition, or "warm-up". Electromyographic studies showed increased insertional activity and high frequency repetitive discharges. Fatty degeneration and pathological changes, including central nuclei, ring fibers, and sarcoplasmic masses, were easily visible in muscle, primarily in type IIA and IIB fibers. Type I fibers appeared unaffected, although this is likely a reflection of differences in avian and human physiology. More importantly, other changes, including reduced fertility, testicular atrophy, and lenticular opacities, were variably present in mutant quails, consistent with the multisystemic nature so characteristic of DM. Although we might expect conservation of DMPK genetic structure and function to extend from mammals to birds, the genetic status of these birds has not been reported. Demonstration of an expanding TNR in this strain or of a genetic defect in a pathway complementary to DMPK function will provide an exciting research opportunity.

5.2
Models of Trinucleotide Expansion

Many laboratories are currently engaged in the establishment of transgenic mice bearing unstable TNR. If the highly directional, dynamic expansion process seen for human TNR disorders when repeats exceed the length threshold can be mimicked, the study of the molecular mechanisms involved in repeat expansion will be greatly facilitated. Perhaps more relevant to this review, mice bearing expanded repeats might exhibit elements of DM pathology not seen in the current mouse models. Strategies could then be devised to dissect the contributions of expanded DMPK transcripts from alterations in protein expression levels. Two studies have recently established arrays of CTG repeats in mice and demonstrated intergenerational instability. In one, a 45 kb genomic cosmid obtained from a mildly affected patient with a 55 repeat allele was used (Gourdon et al. 1997). This fragment also contained the upstream and downstream

DMR-N9/59 and DMAHP genes. Six of seven lines and 6.8 % of all progeny had changes in repeat length. The majority of the changes (11 out of 14) consisted of a gain of a single repeat; single examples of -1, +2, and +6 repeats were observed. In the second study, a fragment of the 3'-UTR contained 162 CTG repeats was used (Monckton et al. 1997). Mutation rates from transmitting females were fairly consistent (approximately 60 %) in all lines studied, and the changes involved were primarily deletions. In contrast, male mutation rates were variable in different lines (9–68 %) and biased towards expansion. Most gains were of one to two repeats, with the largest observed being seven. In addition, reduced litter size was observed in one line, and significant segregation distortion in favor of the transgene in two lines. The instability observed in these studies is modest in comparison to similarly sized repeats in patients, and no dramatic increases were noted. However, different stability thresholds may operate in the mouse. The paternal bias observed by Monckton et al. (1997) and the demonstration of somatic instability by Gourdon et al. (1997) are both encouraging in terms of the dynamic behavior of DM CTG repeats. In particular, sperm from one offspring showing a somatic gain of one repeat exhibited more extreme mosaicism, with additional size gains of up to six repeats visible (Gourdon et al. 1997). This bodes well for continued generational expansion. However, it must again be noted that the modest instability of human CTG repeats in the normal range may be distinct at some molecular level from the dramatic gains responsible for anticipation in DM pedigrees. Meeting this more stringent definition of instability will be necessary before studies relevant to DM can be performed.

6
Concluding Remarks

Studies over the part 5 years have largely delineated the genetic determinants underlying DM transmission patterns, disease severity, and the evolution and maintenance of mutations. In many ways, the opportunity to compare and contrast DM with the other TNR disorders has greatly facilitated progress in this area. In addition, the behavior of very large expansions can be studied in the case of DM and FRAXA, as repeats are more easily tolerated when situated in the UTR of their respective transcripts. This should facilitate the determination of the mechanism of repeat expansion. At the other end of the "expansion spectrum," CAG coding region repeats appear to reach a threshold instability at a slightly smaller size, and small changes in repeat number have dramatic consequences in terms of the health of affected individuals. This should eventually aid in determining the function of the respective protein products, as a gain of function is almost certainly involved in disease.

At this stage, however, the etiology of DM remains far from clear. Research has therefore now reached a critical stage in which the pathological consequences of CTG expansions are being grappled with. In particular, a consensus has not yet been reached on the fundamental questions of whether DM is a single- or multigene disorder and whether decreased levels of DMPK are an important element in disease. Continued analysis of the processing and localization of DMPK transcripts, analysis of the expression of neighboring genes, and the availability of specific antisera should allow the resolution of these issues in the near future. Furthermore, the identification of the substrate or substrates and metabolic role of DMPK will provide another front on which the disturbances common to affected tissues can be analyzed. It has often been speculated that a distinct or additional mechanism operates in cDM over adult-onset disease, due to its unique features. Perhaps an irregularity in transcript metabolism and insufficiency of DMPK both operate when DM repeats become very long. Animal models will hopefully provide a complete organism for testing hypotheses, but much refinement to current approaches appears to be necessary. This resource may already be available in the LWC quail, which should therefore be pursued. As vigorous research utilizing all of these diverse approaches is currently underway, a unified model of the basis of DM appears feasible, ultimately allowing the design of therapeutic approaches. In addition, interesting information on such diverse topics as muscle development, transcript metabolism, and kinase function will undoubtedly be gained during this endeavor.

References

Almqvist E, Spence N, Nichol K, Andrew SE, Vesa J, Peltonen L, Anvret M, Goto J, Kanazawa I, Goldberg YP, Hayden MR (1995) Ancestral differences in the distribution of the Δ2642 glutamic acid polymorphism is associated with varying CAG repeat lengths on normal chromosomes: insights into the genetic evolution of Huntington disease. Hum Mol Genet 4:207–214

Andrew SE, Goldberg YP, Theilmann J, Ziesler J, Hayden MR (1994) A CCG repeat polymorphism adjacent to the CAG repeat in the Huntington disease gene, implications for diagnostic accuracy and predictive testing. Hum Mol Genet 3:65–67

Anvret M, Åhlberg G, Grandell U, Hedberg B, Johnson K, Edström L (1993) Larger expansion of the CTG repeat in muscle compared to lymphocytes from patients with myotonic dystrophy. Hum Mol Genet 2:1397–1400

Argos P, Rao JK (1986) Prediction of protein structure. Methods Enzymol 130:185–207

Ashizawa T, Epstein HF (1991) Ethnic distribution of myotonic dystrophy gene. Lancet 338:642–643

Ashizawa T, Dunne CJ, Dubel JR, Perryman MB, Epstein HF, Boerwinkle E, Hejtmancik JF (1992a) Anticipation in myotonic dystrophy. I. Statistical verification based on clinical and haplotype findings. Neurology 42:1871–1877

Ashizawa T, Dubel JR, Dunne PW, Dunne CJ, Fu Y-H, Pizzuti A, Caskey CT, Boerwinkle E, Perryman MB, Epstein HF, Hejtmancik JF (1992b) Anticipation in myotonic dystrophy. II. Complex relationships between clinical findings and structure of the GCT repeat. Neurology 42:1877–1883

Ashizawa T, Dubel JR, Harati Y (1993) Somatic instability of CTG repeat in myotonic dystrophy. Neurology 43:2674–2678

Ashizawa T, Dunne PW, Ward PA, Seltzer WK, Richards CS (1994a) Effects of the sex of myotonic dystrophy patients on the unstable triplet repeat in their affected offspring. Neurology 44:120–122

Ashizawa T, Anvret M, Baiget M, Barceló JM, Brunner H, Cobo AM, Dallapiccola B, Fenwick RG Jr, Grandell U, Harley H, Junien C, Koch MC, Korneluk RG, Lavedan C, Miki T, Mulley JC, López de Munain A, Novelli G, Roses AD, Seltzer WK, Shaw DJ, Smeets H, Sutherland GR, Yamagata H, Harper PS (1994b) Characteristics of intergenerational contractions of the CTG repeat in myotonic dystrophy. Am J Hum Genet 54:414–423

Ashley CT Jr., Warren ST (1995) Trinucleotide repeat expansion and human disease. Annu Rev Hum Genet 29:703–728

Aslanidis C, Jansen G, Amemiya C, Shutler G, Mahadevan M, Tsilifidis C, Chen C, Alleman J, Wormskamp NGM, Vooijs M, Buxton J, Johnson K, Smeets HJM, Lennon GG, Carrano AV, Korneluk RG, Wieringa B, de Jong PJ (1992) Cloning of the essential myotonic dystrophy region and mapping of the putative defect. Nature (Lond) 355:548–551

Bachner D, Manca P, Steinbach D, Wohrle W, Just W, Vogel W, Hameistere H, Poustka A (1993) Enhanced expression of the murine FMR1 gene during germ cell proliferation suggests a special function in both the male and female gonad. Hum Mol Genet 2:2043–2050

Barceló JM, Mahadevan MS, Tsilfidis C, MacKenzie AE, Korneluk RG (1993) Intergenerational stability of the myotonic dystrophy protomutation. Hum Mol Genet 2:705–709

Bell J (1947) Dystrophia myotonica and allied diseases. In: Treasury of human inheritance, vol 4. Cambridge University Press, Cambridge

Bergoffen J, Kant J, Sladky J, McDonald-McGinn D, Zackai EH, Fischbeck KH (1994) Paternal transmission of congenital myotonic dystrophy. J Med Genet 31:518–520

Bhagwati S, Ghatpande A, Leung B (1996a) Normal levels of DM RNA and myotonin protein kinase in skeletal muscle from adult myotonic dystrophy (DM) patients. Biochim Biophys Acta 1317:155–157

Bhagwati S, Ghatpande A, Leung B (1996b) Identification of two nuclear proteins which bind to RNA CUG repeats: significance for myotonic dystrophy. Biochem Biophys Res Commun 228:55–62

Bouchard G, Roy R, Declos M, Mathieu J, Kouladjian K (1989) Origin and diffusion of the myotonic dystrophy gene in the Saguenay region. Can J Neurol Sci 116:119–122

Boucher CA, King SK, Carey N, Krahe R, Winchester CL, Rahman S, Creavin T, Meghji P, Bailey MES, Chartier FL, Brown SD, Siciliano MJ, Johnson KJ (1995) A novel homeodomain-encoding gene is associated with a large CpG island interrupted by the myotonic dystrophy unstable $(CTG)_n$ repeat. Hum Mol Genet 4:1919–1925

Braga III IS, Oda K, Kikuchi T, Tanaka S, Shin Y, Sento M, Itakura C, Mizutani M (1995) A new inherited muscular disorder in Japanese quails (Coturnix coturnix japonica). Vet Pathol 32:351–360

Brewster BS, Jeal S, Strong PN (1993) Identification of a protein product of the myotonic dystrophy gene using peptide specific antibodies. Biochem Biophys Res Commun 194:1256–1260

Brook JD, McCurrach ME, Harley HG, Buckler AJ, Church D, Aburatani H, Hunter K, Stanton VP, Thirion J-P, Hudson T, Sohn R, Zemelman B, Snell RG, Rundle SA, Crow S, Davies J, Shelbourne P, Buxton J, Jones C, Juvonen, V, Johnson K, Harper PS, Shaw DJ, Housman DE (1992) Molecular basis of myotonic dystrophy: expansion of a trinucleotide (CTG) repeat at the 3' end of a transcript encoding a protein kinase family member. Cell 68:799–808

Brunner HG, Brüggewirth HT, Nillesen W, Jansen G, Hamel BCJ, Hoppe RLE, de Die CEM, Höweler CJ, van Oost BA, Wieringa B, Ropers HH, Smeets HJM (1993a) Influence of sex of the transmitting parent as well as of parental allele size on the CTG expansion in myotonic dystrophy (DM). Am J Hum Genet 53:1016–1023

Brunner HG, Jansen G, Nillesen W, Nelen MR, de Die CEM, Höweler CJ, van Oost BA, Wieringa, B, Ropers H-H, Smeets HJM (1993b) Reverse mutation in myotonic dystrophy. N Engl J Med 328:476–480

Bush EW, Taft CS, Meixell GE, Perryman MB (1996) Overexpression of myotonic dystrophy kinase in BC_3H1 cells induces the skeletal muscle phenotype. J Biol Chem 271:548–552

Buxton J, Shelbourne P, Davies J, Jones C, van Tongeren T, Aslanidis C, de Jong P, Jansen G, Anvret M, Riley B, Williamson R, Johnson K (1992) Detection of an unstable fragment of DNA specific to individuals with myotonic dystrophy. Nature (Lond) 355:547–548

Campuzano V, Montermini L, Molto MD, Pianese L, Cossée M, Cavalcanti F, Monros E, Rodius F, Duclos F, Monticelli A, Zara F, Cañizares J, Koutnikova H, Bidichandani SI, Gellera C, Brice A, Trouillas P, de Michele G, Filla A, de Frutos R, Palau F, Patel PI, di Donato S, Mandel J-L, Cocozza S, Koenig M, Pandolofo M (1996) Friedreich's Ataxia: Autosomal recessive disease caused by an intronic GAA triplet expansion. Science 271:1423–1427

Carango P, Noble JE, Marks HG, Funanage VL (1993) Absence of myotonic dystrophy protein kinase (DMPK) mRNA as a result of a triplet repeat expansion in myotonic dystrophy. Genomics 18:340–348

Carey N, Johnson K, Nokelainen P, Peltonen L, Savontaus M-L, Juvonen V, Anvret M, Grandell U, Chotai K, Robertson E, Middleton-Price H, Malcolm S (1994) Meiotic drive at the myotonic dystrophy locus? Nat Genet 6:117–118

Chakraborty R, Stivers DN, Deka R, Yu LM, Shriver MD, Ferrell RE (1996) Segregation distortion of the CTG repeats at the myotonic dystrophy locus. Am J Hum Genet 59:109–118

Cobo A, Grinberg D, Balcells S, Vilageliu L, Duarte-Gonzalez R, Baiget M (1992) Linkage disequilibrium detected between myotonic dystrophy and the anonymous marker D19S63 in the Spanish population. Hum Genet 89:287–291

Cobo AM, Baiget M, López de Munain A, Poza JJ, Emparanza JI, Johnson K (1993) Sex-related difference in intergenerational expansion of myotonic dystrophy gene. Lancet 341:1159–1160

Cobo AM, Poza JJ, Martorell L, López de Munain A, Emparanza JI, Baiget M (1995) Contribution of molecular analyses to the estimation of the risk of congenital myotonic dystrophy. J Med Genet 32:105–108

Daniels R, Kinis T, Serhal P, Monk M (1995) Expression of the myotonin protein kinase gene in preimplantation human embryos. Hum Mol Genet 4:389–393

Davies J, Yamagata H, Shelbourne P, Buxton J, Ogihara T, Nokelainen P, Nakagawa M, Williamson R, Johnson K, Miki T (1992) Comparison of the myotonic dystrophy associated CTG repeat in European and Japanese populations. J Med Genet 29:766–769

de Die-Smulders CEM, Höweler CJ, Mirandolle JF, Brunner HG, Hovers V, Brüggenwirth H, Smeets HJM, Geraedts JPM (1994) Anticipation resulting in the elimination of the myotonic dystrophy gene: a follow up study of one extended family. J Med Genet 31:595–601

Deka R, Majumder PP, Shriver MD, Stivers DN, Zhong YX, Yu LM, Barrantes R, Yin SJ, Miki T, Hundrieser J, Bunker CH, McGarvey ST, Sakallah S, Ferrell RE, Chakraborty R (1996) Distribution and evolution of the CTG repeats at the myotonin protein kinase gene in human populations. Genome Res 6:142–154

Dunne PW, Walch ET, Epstein HF (1994) Phosphorylation reactions of recombinant human myotonic dystrophy protein kinase and their inhibition. Biochemistry 33:10809–10814

Dunne PW, Ma L, Casey DL, Harati Y, Epstein HF (1996a) Localization of myotonic dystrophy protein kinase in skeletal muscle and its alteration with disease. Cell Motil Cytoskel 33:52–63

Dunne PW, Ma L, Casey DL, Epstein HF (1996b) Myotonic protein kinase expression in human and bovine lenses. Biochem Biophys Res Commun 225:281–288

Edwards A, Hammond HA, Jin L, Caskey CT, Chakraborty R (1992) Genetic variation at five trimeric and tetrameric tandem repeat loci in four human population groups. Genomics 12:241–253

Etongué-Mayer P, Faure R, Bouchard J-P, Thibault M-C, Puymirat J (1994) The myotonin-protein kinase phosphorylates tyrosine residues in normal human skeletal muscle. Biochem Biophys Res Commun 199:89–92

Farkas-Bargeton E, Barbet JP, Dancea S, Wehrle R, Checouri A, Dulac O (1988) Immaturity of muscle fibres in the congenital form of myotonic dystrophy: its consequences and its origin. J Neurol Sci 83:145–159

Fischbeck KH (1994) The mechanism of myotonic dystrophy. Ann Neurol 35:255–256

Fu Y-H, Kuhl D, Pizzuti A, Pierretti M, Sutcliffe JS, Richards S, Verkerk AJMH, Holden JJA, Fenwick RG Jr., Warren ST, Oostra BA, Nelson DL, Caskey CT (1991) Variation of the CGG repeat at the fragile X site results in genetic instability: resolution of the Sherman paradox. Cell 67:1047–1058.

Fu Y-H, Pizzuti A, Fenwick RG, King J Jr., Rajnarayan S, Dunne PW, Dubel J, Nasser GA, Ashizawa T, de Jong PJ, Wieringa B, Korneluk R, Perryman MB, Epstein HF, Caskey CT (1992) An unstable triplet repeat in a gene related to myotonic dystrophy. Science 255:1256–1258

Fu Y-H, Friedman DL, Richards S, Pearlman JA, Gibbs RA, Pizzuti A, Ashizawa T, Perryman MB, Scarlato G, Fenwick RG Jr., Caskey CT (1993) Decreased expression on myotonin-protein kinase messenger RNA and protein in adult form of myotonic dystrophy. Science 260:235–238

Funanage VL, Singleton KS, Carango P, Moses PA, Markes HG (1996) Inverse correlation between number of CTG repeats in the 3' untranslated region of the myotonic dystrophy protein kinase (DMPK) gene and chromatin sensitivity of this region. Am J Hum Genet 59(Suppl):A259

Gacy AM, Goellner G, Juranic N, Macura S, McMurray CT (1995) Trinucleotide repeats that expand in human disease form hairpin structures in vitro. Cell 81:533–540

Gennarelli M, Dallapiccola B, Baiget M, Martorell L, Novelli G (1994) Meiotic drive at the myotonic dystrophy locus. J Med Genet 31:980

Gennarelli M, Lucarelli M, Zelano G, Pizzuti A, Novelli G, Dallapiccola B (1995) Differant expression of the myotonin protein kinase gene in discrete areas of the brain. Biochem Biophys Res Commun 216:489–494

Gibbs M, Collick A, Kelly RG, Jeffreys A (1993) A tetranucleotide repeat mouse minisatellite displaying substantial somatic instability during early preimplantation development. Genomics 17:121–128

Giordano M, De Angelis MS, Mutani R, Richiardi PM (1994) Origin of a regressed myotonic dystrophy allele. J Med Genet 31:130–132

Goldberg YP, Kremer B, Andrew SE, Theilmann J, Graham RK, Squitieri F, Telenius H, Adam S, Sajoo A, Starr E, Heiberg A, Wolff G, Hayden MR (1993) Molecular analysis of new mutations for Huntington's disease: intermediate alleles and sex of origin effects. Nat Genet 5:174–179

Goldman A, Ramsay M, Jenkins T (1994) Absence of myotonic dystrophy in southern African negroids is associated with a significantly lower number of CTG trinucleotide repeats. J Med Genet 31:37–40

Goldman A, Ramsay M, Jenkins T (1995) New founder haplotypes at the myotonic dystrophy locus in Southern Africa. Am J Hum Genet 56:1373–1378

Goldman A, Krause A, Ramsay M, Jenkins T (1996) Founder effect and the prevalence of myotonic dystrophy in South Africans: molecular studies. Am J Hum Genet 59:445–452

Gourdon G, Radvanyi F, Lia A-S, Duros C, Blanche M, Abitbol M, Junien C, Hoffman-Radvanyi H (1997) Moderate intergenerational and somatic instability of a 55-CTG repeat in transgenic mice. Nat Genet 15:190–192

Hamshere MG, Brook JD (1996) Myotonic dystrophy, knockouts, warts and all. Trends Genet 12:332–334

Hamshere M, Newman E, Alwazzan M, Brook JD (1996) Nuclear retention of DMPK transcripts in myotonic dystrophy. Am J Hum Genet 59(Suppl):A262

Hansen RS, Canfield TK, Lamb MM, Gartier SM, Laird CD (1993) Association of fragile X syndrome with delayed replication of the FMR1 gene. Cell 73:1403–1409

Harley HG, Brook JD, Floyd J, Rundle SA, Crow S, Walsh KV, Thibault M-C, Harper PS, Shaw DJ (1991) Detection of linkage disequilibriuim between the myotonic dystrophy locus and a new polymorphic DNA marker. Am J Hum Genet 49:68–75

Harley HG, Brook JD, Rundle SA, Crow S, Reardon W, Buckler AJ, Harper PS, Housman DE, Shaw DJ (1992) Expansion of an unstable DNA region and phenotypic variation in myotonic dystrophy. Nature (Lond) 355:545–546

Harley HG, Rundle SA, MacMillan JC, Myring J, Brook JD, Crow S, Reardon W, Fenton I, Shaw DJ, Harper PS (1993) Size of the unstable CTG repeat sequence in realtion to phenotype and parental transmission in myotonic dystrophy. Am J Hum Genet 52:1164–1174

Harper PS (1989) Myotonic dystrophy, 2nd edn. WB Saunders, London

Harper PS, Dyken PR (1972) Early onset dystrophia myotonica-evidence supporting a maternal environmental factor. Lancet 2:53–55

Harris S, Moncrieff C, Johnson K (1996) Myotonic dystrophy: will the real gene please step forward! Hum Mol Genet 5:1417–1423

Hecht BK, Donnelly A, Gedeon AK, Byard RW, Haan EA, Mulley JC (1993) Direct molecular diagnosis of myotonic dystrophy. Clin Genet 43:276-285

Heitz D, Devys D, Imbert G, Kretz C, Mandel J-L (1992) Inheritance of the fragile X syndrome: size of the fragile X premutation is a major determinant of the transition to full mutation. J Med Genet 29:794-801

Hockings GI, Grice JE, Crosbie GV, Walters MM, Jackson RV (1993) Altered hypothalamic-pituitary-adrenal axis responsiveness in myotonic dystrophy: in vivo evidence for abnormal dihydropyridine-insensitive calcium transport. J Clin Endocrinol Metab 76:1433-1438

Hoffmann-Radvanyi H, Lavedan C, Rabès J-P, Savoy D, Duros C, Johnson K, Junien C (1993) Myotonic dystrophy: absence of CTG enlarged transcript in congenital forms, and low expression of the normal allele. Hum Mol Genet 2:1263-1266

Höweler CJ, Busch HFM, Geraedts JPM, Niermeijer MF, Staal A (1989). Anticipation in Myotonic Dystrophy: fact or fiction? Brain 112:779-797

Hunter A, Tsilfidis C, Mettler G, Jacob P, Mahadevan M, Surh L, Korneluk R (1992) The correlation of age of onset with CTG trinucleotide repeat amplification in myotonic dystrophy. J Med Genet 29:774-779

Hurst GDD, Hurst LD, Barrett JA (1995) Meiotic drive and myotonic dystrophy. Nat Genet 10:132-133

Ikeuchi T, Igarashi S, Takiyama Y, Onodera O, Oyake M, Takano H, Koide R, Tanaka H, Tsuji S (1996) Non-Mendelian transmission in dentatorubral-pallidoluysian atrophy and Machado-Joseph disease: the mutant allele is preferentially tranmitted in male meiosis. Am J Hum Genet 58:730-733

Imbert G, Kretz C, Johnson K, Mandel J-L (1993) Origin of the expansion mutation in myotonic dystrophy. Nat Genet 4:72-76

Imbert G, Saudou F, Yvert G, Devys D, Trottier Y, Garnier J-M, Weber C, Mandel J-L, Cancel G, Abbas N, Durr A, Didierjean O, Stevanin G, Agid Y, Brice A (1996) Cloning of the gene for spinocerebellar ataxia 2 reveals a locus with high sensitivity to expanded CAG/glutamine repeats. Nat Genet 14:285-291

Jacobs AEM, Benders AAGM, Oosterhof A, Veerkamp JH, van Mier P, Wevers RA, Joosten EMG (1991) The calcium homeostasis and the membrane potential of cultured muscle cells from patients with myotonic dystrophy. Biochim Biophys Acta 1096:14-19

Jansen G, Mahadevan M, Amemiya C, Wormskamp N, Segers B, Hendriks W, O'Hoy K, Baird S, Sabourin L, Lennon G, Jap PL, Iles D, Coerwinkel M, Hofker M, Carrano AV, de Jong PJ, Korneluk RG, Wieringa B (1992) Characterization of the myotonic dystrophy region predicts multiple isoform-encoding mRNAs. Nat Genet 1:261-268

Jansen G, Bartolomei M, Kalscheuer V, Merkx G, Wormskamp N, Mariman E, Smeets D, Ropers H-H, Wieringa B (1993) No imprinting involved in the expression of DM-kinase mRNAs in mouse and human tissues. Hum Mol Genet 2: 1221-1227

Jansen G, Willems P, Coerwinkel M, Nillesen W, Smeets H, Vits L, Höweler C, Brunner H, Wieringa B (1994) Gonosomal mosaicism in myotonic dystrophy patients: involvement of mitotic events in $(CTG)_n$ repeat variation and selection against extreme expansion in sperm. Am J Hum Genet 54:575-585

Jansen G, Bächner D, Coerwinkel M, Wormskamp N, Hameister H, Wieringa B (1995) Structural organization and developmental expression pattern of the mouse WD-repeat gene DMR-N9 immediately upstream of the myotonic dystrophy locus. Hum Mol Genet 4:843-852

Jansen G, Groenen PJTA, Bächner D, Jap PHK, Coerwinkel M, Oerlemans F, van den Broek W, Gohlsch B, Pette D, Plomb JJ, Molenaar FC, Nederhoff MGJ, van Echteld CJA, Dekker M, Berns A, Hameister H, Wieringa B (1996) Abnormal myotonic dystrophy protein kinase levels produce only mild myopathy in mice. Nat Genet 13:316-324

Jones C, Penny L, Mattina T, Yu S, Baker E, Voullaire L, Langdon WY, Sutherland GR, Richards RI, Tunnacliffe A (1995) Association of a chromosome deletion syndrome with a fragile site within the protooncogene CBL2. Nature (Lond) 376:145-149

Justice RW, Zilian O, Woods DF, Noll M, Bryant PJ (1995) The Drosophilia tumor suppressor gene warts encodes a homolog of human myotonic dystrophy kinase and is required for the control of cell shape and proliferatrion. Genes Dev 9:534-546

Kawaguchi Y, Okamoto T, Taniwaki M, Aizawa M, Inoue M, Katayama S, Kawakami H, Nakamura S, Nishimura M, Akiguchi I, Kimura J, Narumiya S, Kakizuka A (1994) CAG expansions in a novel gene for Machado-Joseph disease at chromosome 14q32.1. Nat Genet 8:221–227

Kawakami K, Ohto H, Ikeda K, Roeder RG (1996) Structure, function and expression of a murine homeobox protein AREC3, a homologue of Drosophila sine oculis gene product, and implication in development. Nucleic Acids Res 24:303–310

Khandjian EW, Corbin F, Woerly S, Rousseau F (1996) The fragile X mental retardation protein is associated with ribosomes. Nature (Lond) 12:91–93

King SK, Wells DJ, Wells KE, Carey N, Johnson KJ (1996) A 3.7 kb fragment from the myotonic dystrophy protein kinase directs neural-specific expression in vivo. Biochem Soc Trans 24:283S

Kinoshita M, Takahashi R, Hasegawa T, Komori T, Nagasawa R, Hirose K, Tanabe H (1996) $(CTG)_n$ expansions in various tissues from a myotonic dystrophy patient. MuscleNerve 19:240–242

Koch MC, Grimm T, Harley HG, Harper PS (1991) Genetic risks for children of women with myotonic dystrophy. Am J Hum Genet 48:1084–1091

Koga R, Nakao Y, Kurano Y, Tsukahara T, Nakamura A, Ishiura S, Nonaka I, Arahata K (1994) Decreased myotonin-protein kinase in the skeletal and cardiac muscles in myotonic dystrophy. Biochem Biophys Res Commun 202:577–585

Krahe R, Eckhart M, Ogunniyi AO, Osuntokun BO, Siciliano MJ, Ashizawa T (1995a) De novo myotonic dystrophy mutation in a Nigerian kindred. Am J Hum Genet 56:1067–1074

Krahe R, Ashizawa T, Abbruzzese C, Roeder E, Carango P, Giacanelli M, Funanage VL, Siciliano MJ (1995b) Effect of myotonic dystrophy trinucleotide repeat expansion on DMPK transcription and processing. Genomics 28:1–14

Kunst CB, Warren ST (1994) Cryptic and polar variation of the fragile X repeat could result in predisposing normal alleles. Cell 77:853–861

La Spada AR, Wilson EM, Lubahn DB, Harding AE, Fischbeck KH (1991) Androgen receptor gene mutations in X-linked spinal and bulbar muscular atrophy. Nature (Lond) 352:77–79

La Spada AR, Roling DB, Harding AE, Warner CL, Spiegel R, Hausmanowa-Petrusewicz I, Yee W-C, Fischbeck KH (1992) Meiotic stability and genotype-phenotype correlation of the trinucleotide repeat in the X-linked spinal and bulbar muscular atrophy. Nat Genet 2:301–304

Lavedan C, Hoffman-Radvanyi H, Shelbourne P, Rabes J-P, Duros C, Savoy D, Dehaupas I, Luce S, Johnson K, Junien C (1993a) Myotonic dystrophy: size- and sex-dependent dynamics of CTG meiotic instability, and somatic mosaicism. Am J Hum Genet 52:875–883

Lavedan C, Hofmann-Radvanyi H, Rabes JP, Roume J, Junien C (1993b) Differant sex-dependent constraints in CTG length variation as explanation for congenital myotonic dystrophy. Lancet 341:237

Lavedan C, Hoffman-Radvanyi H, Boileau C, Bonaïti-Pellié C, Savoy D, Shelbourne P, Duros C, Rabes J-P, Dehaupas I, Luce S, Johnson K, Junien C (1994) French myotonic dystrophy families show expansion of a CTG repeat in complete linkage disequilibrium with an intragenic 1 kb expansion. J Med Genet 31:33–36

Leeflang EP, Arnheim N (1995) A novel repeat structure at the myotonic dystrophy locus in a 37 repeat allele with unexpected high stability. Hum Mol Genet 4:135–136

Leeflang EP, Zhang L, Tavaré S, Hubert R, Srinidhi J, MacDonald ME, Myers RH, de Young M, Wexler NS, Gusella JF, Arnheim N (1995) Single sperm analysis of the trinucleotide repeats in the Huntington's disease gene: quantification of the mutation frequency spectrum. Hum Mol Genet 4:1519–1526

Leeflang EP, McPeek MS, Arnheim N (1996) Analysis of mitotic segregation, using single-sperm typing: meiotic drive at the myotonic dystrophy locus. Am J Hum Genet 59:896–904

Lim L, Hall C, Monfries C (1996) Regulation of actin cytoskeleton by rho-family GTPases and their associated proteins. Semin Cell Dev Biol 7:699–706

López de Munain A, Cobo AM, Huguet E, Marri Massó JF, Johnson K, Baiget M (1994) CTG trinucleotide repeat variability in identical twins with myotonic dystrophy. Ann Neurol 35:374–375

Lyttle TW (1993) Cheaters sometimes prosper: distortion of mendelian segregation by meiotic drive. Trends Genet 9:205–210

MacDonald ME, Novelletto A, Lin C, Tagle D, Barnes G, Bates G, Taylor S, Allitto B, Altherr M, Myers R, Lehrach H, Collins FS, Wasmuth JJ, Frontali M, Gusella JF (1992) The Huntington's disease candidate region exhibits many different haplotypes. Nat Genet 1:99–103

MacKenzie AE, Macleod HL, Korneluk RG (1989) Linkage analysis of the apolipoprotein C2 gene and myotonic dystrophy on human chromosome 19 reveals linkage disequilibrium in a French-Canadian population. Am J Hum Genet 44:140–147

Maeda M, Taft CS, Bush EW, Holder E, Bailey WM, Neville H, Perryman MB, Bies RD (1995) Identification, tissue-specific expression, and subcellular localization of the 80- and 71-kDa forms of myotonic dystrophy kinase protein. J Biol Chem 270:20246–20249.

Mahadevan M, Tsilfidis C, Sabourin L, Shutler G, Amemiya C, Jansen G, Neville C, Narang M, Barceló J, O'Hoy K, Leblond S, Earle-MacDonald J, de Jong P.J, Wieringa B, Korneluk RG (1992) Myotonic dystrophy mutation: an unstable CTG repeat in the 3' untranslated region of the gene. Science 255:1253–1255

Mahadevan MS, Amemiya C, Jansen G, Sabourin L, Baird S, Neville CE, Wormskamp N, Segers B, Batzer M, Lamerdin J, de Jong P, Wieringa B, Korneluk RG (1993a) Structure and genomic sequence of the myotonic dystrophy (DM kinase) gene. Hum Mol Genet 2:299–304

Mahadevan MS, Foitzik MA, Surh LC, Korneluk RG (1993b) Characterization and polymerase chain reaction (PCR) detection of an *Alu* polymorphism in total linkage disequilibrium with myotonic dystrophy. Genomics 15:446–448

Martorell L, Martinez JM, Carey N, Johnson K, Baiget M (1995) Comparison of CTG repeat length expansion and clinical progression of myotonic dystrophy over a five year period. J Med Genet 32:593–596

Monckton DG, Wong L-JC, Ashizawa T, Caskey CT (1995) Somatic mosaicism, germline expansions, germline reversions and intergenerational reductions in myotonic dystrophy males: small pool PCR analyses. Hum Mol Genet 4:1–8

Monckton DG, Coolbaugh MI, Ashizawa KT, Siciliano MJ, Caskey CT (1997) Hypermutable myotonic dystrophy CTG repeats in transgenic mice. Nat Genet 15:193–196

Mounsey JP, Xu P, John III JE, Horne LT, Gilbert J, Roses AD, Moorman JR (1995) Modulation of skeletal muscle sodium channels by human myotonin protein kinase. J Clin Invest 95:2379–2384

Mulley JC, Staples A, Donnelly A, Gedeon AK, Hecht BK, Nicholoson GA, Haan EA, Sutherland GR (1993) Explanation for exclusive maternal origin for congenital form of myotonic dystrophy. Lancet 341:236–237

Nakagawa M, Yamada H, Higuchi I, Kamanishi Y, Miki T, Johnson K, Osame M (1994) A case of paternally inherited congenital myotonic dystrophy. J Med Genet 31:397–400

Neville CE, Mahadevan MS, Barceló JM, Korneluk RG (1994) High resolution genetic analysis suggests one ancestral predisposing haplotype for the origin of the myotonic dystrophy mutation. Hum Mol Genet 3:45–51

Nokelainen P, Alanen-Kurki L, Winqvist R, Falck B, Somer H, Leisti J, Savontaus ML, Peltonen L (1990) Linkage disequilibrium detected between dystrophia myotonica and APOC2 locus in the Finnish population. Hum Genet 85:541–545

Norbury G, Kiernan E, Miciak A (1993) Analysis of the expansion associated with myotonic dystrophy in various tissues from fetuses predicted to be at high risk. J Med Genet 30:341

Novelli G, Gennarelli M, Zelano G, Pizzuti A. Fattorini C, Caskey CT, Dallapiccola B (1993) Failure in detecting mRNA transcripts from the mutated allele in myotonic dystrophy muscle. Biochem Mol Biol Int 29:291–297

O'Hoy KL, Tsilfidis C, Mahadevan MS, Neville CE, Barceló J, Hunter AGW, Korneluk RG (1993) Reduction in size of the myotonic dystrophy trinucleotide repeat mutation during transmission. Science 259:809–812

Ohya K, Tachi N, Chiba S, Sato T, Kon S, Kikuchi K, Imamura S, Yamagata H, Miki T (1994) Congenital myotonic dystrophy transmitted from an asymptomatic father with a DM-specific gene. Neurology 44:1958–1960

Orr HT, Chung M-Y, Banfi S, Kwaitkowski TJ Jr., Servadio A, Beaudet AL, McCall AE, Duvick LA, Ranum LPW, Zoghbi HY (1993) Expansion of an unstable trinucleotide CAG repeat in spinocerebellar ataxia type 1. Nat Genet 4:221–226.

Otten AD, Tapscott SJ (1995) Triplet repeat expansion in myotonic dystrophy alters the adjacent chromatin structure. Proc Natl Acad Sci USA 92:5465–5469

Oudet C, Mornet E, Serre JL, Thomas F, Lentes-Zengerling S, Kretz C, Deluchat C, Tejada I, Boué J, Boué A, Mandel JL (1993) Linkage disequilibrium between the fragile X mutation and two closely linked CA repeats suggests that fragile X chromosomes are derived from a small number of founder chromosomes. Am J Hum Genet 52:297–304

Passos-Bueno MR, Cerqueira A, Mainzof M, Marie SK, Zatz M (1994) Myotonic dystrophy: genetic, clinical, and molecular analysis of patients from 41 Brazilian families. J Med Genet 32:14–18

Pulst S-M, Nechiporuk A, Nechiporuk T, Gispert S, Chen X-N, Lopes-Cendes I, Pearlman S, Starkman S, Orozco-Diaz G, Lunkes A, DeJong P, Rouleau GA, Auburger G, Korenberg JR, Figueroa C, Sahba S (1996) Moderate expansion of a normally biallelic trinucleotide repeat in spinocerebellar ataxia type 2. Nat Genet 14:268–276

Rastinejad F, Blau HM (1993) Genetic complementation reveals a novel regulatory role for 3' untranslated regions in growth and differentiation. Cell 72:903–917

Reddy S, Smith DBJ, Rich MM, Leferovich JM, Reilly P, Davis BM, Tran K, Rayburn H, Bronson R, Cros D, Balice-Gordon RJ, Housman D (1996) Mice lacking the myotonic dystrophy protein kinase develop a late onset progressive myopathy. Nat Genet 13:325–335

Redman JB, Fenwick RG Jr., Fu Y-H, Pizzuti A, Caskey CT (1993) Relationship between parental trinucleotide GCT repeat length and severity of myotonic dystrophy in offspring. JAMA 269:1960–1965

Reiss AL, Kazazian HH, Krebs CM, McAughan A, Boehm CD, Abrams MT, Nelson DL (1994) Frequency and stability of the fragile X premutation. Hum Mol Genet 3:393–398

Reyniers E, Vits L, De Boulle K, Van Roy B, Van Velzen D, de Graaff E, Verkerk AJMH, Jorens HZJ, Darby JK, Oostra B, Willems PJ (1993) The full mutation in the FMR-1 gene in male fragile X patients is absent in their sperm. Nat Genet 4:143–146

Richards RI, Sutherland GR (1992) Dynamic mutations: a new class of mutations causing human disease. Cell 70:709–712

Richards RI, Holman K, Friend K, Kremer E, Hillen D, Staples A, Brown WT, Goonewardena P, Tarleton J, Schwartz C, Sutherland GR (1992) Evidence of founder chromosomes in fragile X syndrome. Nat Genet 1:257–260

Roses AD (1994) Muscle biochemistry and a genetic study of myotonic dystrophy. Science 264:587

Rubinsztein DC, Leggo J, Barton DE, Ferguson-Smith MA (1993) Site of (CCG) polymorphism in the HD gene. Nat Genet 5:214–215

Rubinsztein DC, Leggo J, Amos W, Barton DE, Ferguson-Smith MA (1994) Myotonic dystrophy CTG repeats and the associated insertion/deletion polymorphism in human and primate populations. Hum Mol Genet 3:2031–2035

Rubinsztein DC, Leggo J, Goodburn S, Barton DE, Ferguson-Smith MA (1995) Haplotype analysis of the Δ2642 and $(CAG)_n$ polymorphisms in the Huntington's disease (HD) gene provides an explanation for the apparent 'founder' HD haplotype. Hum Mol Genet 4:203–206

Sabourin LA, Mahadevan MS, Narang M, Lee DSC, Surh LC, Korneluk RG (1993) Effect of the myotonic dystrophy (DM) mutation on mRNA levels of the DM gene. Nat Genet 4:233–238

Salvatori S, Biral D, Furlan S, Marin O (1994) Identification and localization of the myotonic dystrophy gene product in skeletal and cardiac muscles. Biochem Biophys Res Commun 203:1365–1370

Sanpei K, Takano H, Igarashi S, Sato T, Oyake M, Sasaki H, Wakisaka A, Tashiro K, Ishida Y, Ikeuchi T, Koide R, Saito M, Sato A, Tanaka T, Hanyu S, Takiyama Y, Nishizawa M, Shimizu N, Nomura Y, Segawa M, Iwabuchi K, Eguchi I, Takahashi H, Tsuji S (1996) Identification of the spinocerebellar ataxia type 2 gene using a direct identification of repeat expansion and cloning technique, DIRECT. Nat Genet 14:277–284

Sasagawa N, Sorimachi H, Maruyama K, Arahata K, Ishiura S, Suzuki K (1994) Expression of a novel human myotonin protein kinase (MtPK) cDNA clone which encodes a protein with a thymopoietin-like domain in COS cells. FEBS Lett 351:22–26

Shaw DJ, Chaudhary S, Rundle SA, Crow S, Brook JD, Harper PS, Harley HG (1993a) A study of DNA methylation in myotonic dystrophy. J Med Genet 30:189–192

Shaw DJ, McCurrach M, Rundle SA, Harley HG, Crow SR, Sohn R, Thirion J-P, Hamshere MG, Buckler AJ, Harper PS, Housman DE, Brcok JD (1993b) Genomic organization and transcriptional units at the myotonic dystrophy locus. Genomics 18: 673–679

Shaw AM, Bernetson RA, Phillips MF, Harper PS, Harley HG (1995) Evidence for meiotic drive at the myotonic dystrophy locus. J Med Genet 32:145

Shelbourne P, Winqvist R, Kunert E, Davies J, Leisti J, Thiele H, Bachmann H, Buxton J, Williamson B, Johnson K (1992) Unstable DNA may be responsible for the incomplete penetrance of the myotonic dystrophy phenotype. Hum Mol Genet 1:467–473

Skerjanc IS, Slack RS, McBurney MW (1994) Cellular aggregation enhances myoD-directed skeletal myogenesis in embryonal carcinoma cells. Mol Cell Biol 14:8451–8459

Snow K, Doud LK, Hagerman R, Pergolizzi RG, Erster SH, Thibodeau SN (1993) Analysis of a CGG sequence at the FMR-1 locus in fragile X families and in the general population. Am J. Hum Genet 53:1217–1228

Squitieri F, Andrew SE, Goldberg YP, Kremer B, Spence N, Zeisler J, Nichol K, Theilmann J, Greenberg J, Goto J, Kanazawa I, Vesa J, Peltonen L, Almqvist E, Anvret M, Telenius H, Lin B, Napolitano G, Morgan K, Hayden MR (1994) DNA haplotype analysis of Huntington disease reveals clues to the origins and mechanisms of CAG expansion and reasons for geographic variations of prevalence. Hum Mol Genet 3:2103–2114

Stief A, Winter DM, Stratling WH, Sippel AE (1989) A nuclear DNA attachment element mediates elevated and position-independent gene activity. Nature (Lond) 341:343–345

Tachi N, Kozuka N, Ohya K, Chiba S, Kikuchi K (1995) Expression of myotonic dystrophy protein kinase in biopsied muscles. J Neurol Sci 132:61–64

Tamanini F, Meijer N, Verheij C, Willems PJ, Galjaard H, Oostra BA, Hoogeveen AT (1996) FMRP is associated to the ribosomes via RNA. Hum Mol Genet 5:809–813

Taneja KL, McCurrach M, Schalling M, Housman D, Singer RH (1995) Foci of trinucleotide repeat transcripts in nuclei of myotonic dystrophy cells and tissues. J Cell Biol 128:995–1002

Telenius H, Kremer B, Goldberg YP, Thielmann J, Andrew SE, Zeisler J, Adam S, Greenberg C, Ives EJ, Clarke LA, Hayden MR (1994) Somatic and gonadal mosaicism of the Huntington disease gene CAG repeat in brain and sperm. Nat Genet 6:409–414

Telenius H, Almqvist E, Kremer B, Spence N, Squitieri F, Nichol K, Grandel U, Starr E, Benjamin C, Castaldo I, Calabrese O, Anvret M, Goldberg YP, Hayden MR (1995) Somatic mosaicism in sperm is associated with intergenerational $(CAG)_n$ changes in Huntington disease. Hum Mol Genet 4:189–195

Thibault MC, Mathieu J, Moorjani S, Lescault A, Prevost C, Gaudet D, Morissette J, Laberge C (1989) Myotonic dystrophy: linkage with apolipoprotein E and estimation of the gene carrier status with genetic markers. Can J Neurol Sci 16:134–140

Thornton CA, Griggs RC, Moxley III RT (1994a) Myotonic dystrophy with no trinucleotide repeat expansion. Ann Neurol 35:269–272

Thornton CA, Johnson K, Moxley III RT (1994b) Myotonic dystrophy patients have larger CTG expansions in skeletal muscle than in leukocytes. Ann Neurol 35:104–107

Thornton CA, Wymer JP, Moxley RT (1996) Expression of the DMAHP gene is supressed in cis by the myotonic dystrophy (MTD) CTG repeat expansion. Am J Hum Genet 59(Suppl):A33

Timchenko LT, Caskey CT (1996) Trinucleotide repeat disorders in humans: discussions of mechanisms and medical issues. FASEB J 10:1589–1597

Timchenko L, Nastainczyk W, Schneider T, Patel B, Hofmann F, Caskey CT (1995) Full-length myotonin protein kinase (72 kDa) displays serine kinase activity. Proc Natl Acad Sci USA 92:5366–5370

Timchenko LT, Timchenko NA, Caskey CT, Roberts R (1996a) Novel proteins with binding specificity for DNA CTG repeats and RNA CUG repeats: implications for myotonic dystrophy. Hum Mol Genet 5:115–121

Timchenko LT, Miller JW, Timchenko NA, DeVore DR, Datar KV, Lin L, Roberts R, Caskey CT, Swanson MS (1996b) Identification of a $(CUG)_n$ triplet repeat RNA-binding protein and its expression in myotonic dystrophy. Nucleic Acids Res 24:4407–4414

Tsilfidis C, MacKenzie AE, Mettler G, Barceló J, Korneluk RG (1992) Correlation between CTG trinucleotide repeat length and frequency of severe congenital myotonic dystrophy. Nat Genet 1:192-195

van der Ven PFM, Jansen G, van Kuppevelt THMSM, Perryman MB, Lupa M, Dunne PW, ter Laak HJ, Jap PHK, Veerkamp JH, Epstein HJ, Wieringa B (1993) Myotonic dystrophy kinase is a component of neuromuscular junctions. Hum Mol Genet 2:1889-1894

Vanier TM (1960) Dystrophia myotonica in childhood. Br Med J 2:1284

Wang J, Pegoraro E, Menegazzo E, Gennarelli M, Hoop RC, Angelini C, Hoffman EP (1995) Myotonic dystrophy: evidence for a possible dominant-negative RNA mutation. Hum Mol Genet 4:599-606

Wang Y-H, Amirhaeri S, Kang S, Wells RD, Griffith JD (1994) Preferential nucleosome assembly at DNA triplet repeats from the myotonic dystrophy gene. Science 265:669-671

Wang Y-H, Griffith J (1995) Expanded CTG triplet blocks from the myotonic dystrophy gene create the strongest known natural nucleosome positioning elements. Genomics 25:570-573

Waring JD, Haq R, Tamai K, Sabourin LA, Ikeda J-E, Korneluk RG (1996) Investigation of myotonic dystrophy kinase isoform translocation and membrane association. J Biol Chem 271:15187-15193

Warren ST (1996) The expanding world of trinucleotide repeats. Science 271:1374-1375

Watanabe M, Abe K, Aoki M, Yasuo K, Itoyama Y, Shoji M, Ikeda Y, Iizuki T, Ikeda M, Shizuka M, Mizushima K, Hirai S (1996) Mitotic and meiotic stability of the CAG repeat in the X-linked spinal and bulbar muscular atrophy gene. Clin Genet 50:133-137

Weber Jl (1990) Informativeness of human $(dC\text{-}dA)_n\text{-}(dG\text{-}dT)_n$ polymorphisms. Genomics 7:524-530

Whiting EJ, Waring JD, Tamai K, Somerville MJ, Hincke M, Staines WA, Ikeda J-E, Korneluk RG (1995) Characterization of myotonic dystrophy kinase (DMK) protein in human and rodent muscle and central nervous tissue. Hum Molec Genet 4:1063-1072

Wieringa B (1994) Myotonic dystrophy reviewed: back to the future? Hum Mol Genet 3:1-7

Wissman A, Ingles J, McGhee JD, Mains PE (1997) Caenorhabditis elegans LET-502 is related to rho-binding kinases and human myotonic dystrophy kinase and interacts genetically with a homolog of the regulatory subunit of smooth muscle myosin phosphatase to affect cell shape. Genes Dev 11:409-422

Wöhrle D, Hennig I, Vogel W, Steinbach P (1993) Mitotic stability of fragile X mutations in differentiated cells indicates early post-conceptional trinucleotide repeat expansion. Nat Genet 4:140-142

Wöhrle D, Kennerknecht I, Wolf M, Enders H, Schwemmle S, Steinbach P (1995) Heterogeneity of DM kinase repeat expansion in differant fetal tissues and further expansion during cell proliferation in vitro: evidence for a causal involvement of methyl-directed DNA mismatch repair in triplet repeat stability. Hum Mol Genet 4:1147-1153

Wong L-CJ, Ashizawa T, Monckton DG, Caskey CT, Richards CS (1995) Somatic heterogeneity of the CTG repeat in myotonic dystrophy is age and size dependent. Am J Hum Genet 56:114-122

Yamagata H, Miki T, Ogihara T, Nakagawa M, Higuchi I, Osame M, Shelbourne P, Davies J, Johnson K (1992) Expansion of unstable DNA region in Japanese myotonic dystrophy patients. Lancet 339:692

Yamagata H, Miki T, Sakoda S-I, Yamanaka N, Davies J, Shelbourne P, Kubota Y, Takenaga S, Nakagawa M, Ogihara T, Johnson K (1994) Detection of a premutation in Japanese myotonic dystrophy. Hum Mol Genet 3:819-820

Yamagata H, Miki T, Nakagawa M, Johnson K, Deka R, Ogihara T (1996) Association of CTG repeats and the 1-kb Alu isertion/deletion polymorphism at the myotonic protein kinase gene in the Japanese population suggests a common Eurasian origin of the myotonic dystrophy mutation. Hum Genet 97:145-147

Yarden O, Plamann M, Ebbole DJ, Yanofsky C (1992) cot-1, a gene required for hyphal elongation in Neurospora crassa, encodes a protein kinase. EMBO J 11:2159-2166

Zatz M, Passos-Bueno MR, Cerqueira A, Marie SK, Vainzof M, Pavanello RCM (1995) Analysis of the CTG repeat in skeletal muscle of young and adult myotonic dystrophy patients: when does the expansion occur? Hum Mol Genet 4:401-406

Zerylnick C, Torroni A, Sherman SL, Warren ST (1995) Normal variation at the myotonic dystrophy locus in global human populations. Am J Hum Genet 56:123–130

Zhang L, Leeflang EP, Yu J, Arnheim N (1994) Studying human mutations by sperm typing: instability of CAG trinucleotide repeats in the human androgen receptor gene. Nat Genet 7:531–535

Zhang L, Fischbeck KH, Arnheim N (1995) CAG repeat length variation in sperm from a patient with Kennedy's disease. Hum Mol Genet 4:303–305

Zheng C-J, Byers B, Moolgavkar SH (1993) Allelic instability in mitosis: a unified model for dominant disorders. Proc Natl Acad Sci USA 90:10178–10182

Instabilities of Triplet Repeats: Factors and Mechanisms

Robert D. Wells[1], Albino Bacolla[1], and Richard P. Bowater[2]

1
Genetic Instabilities of Repetitive DNA and Human Diseases

Repetitive DNA sequences are dispersed throughout natural genomes, both within and outside of known coding sequences (Charlesworth et al. 1994; Tautz and Schlotterer 1994). The varied distribution of particular repetitive sequences between different species suggests that they fulfill specific cellular requirements, but direct evidence for what these may be is lacking. However, any cellular functions are likely to be restricted to eukaryotes since long repeating sequences are absent from prokaryotic genomes.

Variations in the unit size and degree of repetition within DNA repeats has led to their division into a number of categories. In this review, we will concentrate on simple repeating DNA sequences, also termed microsatellites; these sequences are tandem (direct) repeats with a high degree of repetition and are 1–5 bp in their unit structure. More complex repeat units, known as minisatellites, are also found in eukaryotes and have been particularly well characterized in humans (Armour and Jeffreys 1992).

Simple repeating sequences have an intrinsic genetic instability, manifested as frequent length changes due to insertions (also referred to as expansions) or deletions of repeat units. The rate of genetic change that occurs in these sequences is related to their copy number, and a mutated product therefore has a different potential for mutation compared to its predecessor; this phenomenon has been termed dynamic mutation (Richards and Sutherland 1992; Sutherland and Richards 1995).

The mechanisms for evolution of such sequences have been much debated (Dover 1995; Hancock 1996). The genetic instability associated with repeating sequences has been proposed to occur via a variety of pathways which are known to modify the genetic material. Several arguments suggest that changes in repeat number reflect DNA polymerase slippage (Levinson and Gutman 1987b; Lustig and Petes 1993; Hancock 1996). The susceptibility of repetitive DNA sequences to slipped-strand mispairing provides a plausible mechanism

[1] Institute of Biosciences and Technology, Texas A&M University, Texas Medical Center, 2121 W. Holcombe Blvd., Houston, Texas 77030–3303, USA
[2] Mutagenesis Labs, Room C49, Imperial Cancer Research Fund, Clare Hall Laboratories, South Mims, Herts. EN6 3LD, UK

for generating unusual DNA conformations (Sinden and Wells 1992; Wells 1996). This topic will be discussed in more detail below.

Slippage of the DNA strands at the replication fork is the favored mechanism to produce changes within simple repeats, but direct evidence that this occurs inside cells has not been obtained. Hence, it is possible that other mechanisms may be involved. Unequal exchange during recombination between homologous DNA sequences (Smith 1973) is an attractive possibility, because it can produce deletions and insertions from one event. Gene conversion has been proposed to explain polymorphisms at minisatellites (Jeffreys et al. 1994), although its relationship to the smaller repeat units of microsatellites is not known. It is clear that the evolution of repetitive sequences is complex, and it is likely that interactions occur between these different pathways.

Recently, genetic instabilities within repetitive DNA sequences have been linked to a variety of human diseases (Krontiris 1995; Sutherland and Richards 1995). These findings have produced a surge in interest across a range of scientific disciplines. Studies into the molecular mechanisms producing genetic instabilities of simple repeating sequences and their relationship to various hereditary disorders are the topic of this review.

1.1
Diseases Associated with Expansion of Triplet Repeat Sequences

Unusual mutation events involving the expansion of specific triplet repeat sequences (TRS) have been linked to a number of human hereditary neuromuscular or neurodegenerative disorders (Caskey et al. 1992; Willems 1994; Ashley and Warren 1995; Paulson and Fischbeck 1996; Warren 1996). Other candidate diseases are being studied, and it is likely that more will be related to this phenomenon. Most of these disorders show the unusual clinical behavior of anticipation, which is defined as the increased severity and/or decreased age of onset of a hereditary disease with progression through a pedigree. Analysis of TRS has identified the molecular basis for this clinical behavior: decreased age of onset or more severe symptoms of disease correlate with longer TRS, and anticipation occurs because genetic instabilities within TRS can produce longer repeats upon transmission to offspring. Other chapters in this book and recent reviews on myotonic dystrophy (Harris et al. 1996), fragile X syndrome (Warren and Ashley 1995) and Huntington's disease (HD) (Nasir et al. 1996) have discussed the relationship of triplet repeat expansion to the clinical features of their respective diseases.

The events associated with expansion of TRS fall into two categories, termed types 1 and 2 (Paulson and Fischbeck 1996; see Tables 1, 2). In type 1 expansions, the TRS is always CTG·CAG and is in a coding segment of the gene. Disorders in this category are essentially confined to the nervous system and comprise HD (Huntington's Disease Collaborative Research Group 1993), denta-

Table 1. Human type 1 diseases associated with expansion of trinucleotide repeats

Type 1 disease
Characteristics:

CAG repeat in translated region of RNA
small expansions produce longer tracts of polyglutamines
specific neurons affected due to toxic gain of function of protein

Disorder	Gene locus	Repeat Length	
		Normal	Disease
Spinal and bulbar muscular atrophy	AR	11–34	38–66
Huntington's disease	Hdh (IT15)	10–35	36–121
Dentatorubral-pallidoluysian atrophy and Haw River syndrome	DRPLA (B37)	7–25	49–75
Spinocerebellar ataxias			
Type 1	SCA1	6–39	41–81
Type 2	SCA2	15–29	35–59
Type 3 (Machado-Joseph disease)	SCA3 (MJD1)	12–37	61–84

Table 2. Human type 2 diseases associated with expansion of trinucleotide repeats

Type 2 disease
Characteristics:

various repeats in untranslated region of RNA
very large expansions can occur
multi-system disorders caused by altered gene expression

Disorder	Gene locus	Repeat sequence	Repeat length				Repeat location
			Normal	Pre-mutation	Proto-mutation	Disease	
Myotonic dystrophy	DMPK	CAG	5–37	–	50–200	200–3000	3'-Untranslated
Fragile X syndrome	FRAXA	CGG	6–52	60–200	–	230–1000	5'-Untranslated
Friedreich's ataxia	X25	GAA	7–22	–	–	200–900	First intron

torubral-pallidoluysian atrophy (DRPLA; Koide et al. 1994; Nagafuchi et al. 1994), spinal and bulbar muscular atrophy (SBMA, also known as Kennedy's disease; La Spada et al. 1991), and the spinocerebellar ataxias type 1 (SCA1; Orr et al. 1993), type 2 (SCA2; Imbert et al. 1996; Pulst et al. 1996; Sanpei et al. 1996), and type 3 (SCA3, also known as Machado-Joseph disease, MJD; Kawaguchi et al. 1994). In these diseases, expansion events occur over a limited range; normal individuals have up to 30–40 repeat lengths which are stable, and afflicted individuals have 35–100 repeats which are genetically unstable (Tables 1, 2). Hence, there is a threshold length above which each sequence is unstable upon genetic transmission.

The net effect of type 1 expansions is to produce proteins with longer tracts of polyglutamines. The proteins associated with each disease are unrelated and probably have different functions. Thus, although the mutation mechanism producing these extended proteins may be similar, pathways producing the disease pathologies are likely to be different. The expansion events may create proteins with a gain of function that is particularly deleterious in neurons. Proteins have been identified that interact with some of the disease polypeptides via their glutamine tracts (Li et al. 1995; Burke et al. 1996). A stretch of glutamine residues may undergo a conformational change once expanded beyond a crucial threshold (Li et al. 1995; Trottier et al. 1995) and may thus alter the strength of any protein-protein interactions. Hence, it is possible that the selective expression of associated proteins provides the tissue specificity for the diseases.

The various type 1 diseases have quite different pathologies, and a more detailed examination of their respective TRS expansions may provide clues to their disease pathways. For example, the identification of the gene associated with SCA2 showed that disease alleles had somewhat shorter lengths of triplet repeats compared to the other type 1 diseases. Thus it seems likely that expansions within the glutamine tract are tolerated to different levels for the various diseases. Comparison of the expansions associated with the various spinocerebellar ataxias supports this idea (Zoghbi 1996) and shows that disease severity and age of onset are determined by factors other than the length of the polyglutamine tract.

Type 2 expansion events are associated with multisystem disorders and have a number of differences compared to type 1 (Table 2); various types of sequences are found, and the TRS is not located within a coding region of the gene. For example, a CGG·CCG repeat is found in the 5'-untranslated region of the gene in fragile X syndrome (Ashley et al. 1993), a CTG·CAG repeat is found in the 3'-untranslated region of a gene linked with myotonic dystrophy (Fu et al. 1992; Mahadevan et al. 1992), and an AAG·CTT repeat occurs in the first intron of the gene associated with Friedreich's ataxia (Campuzano et al. 1996). Extremely large expansions (hundreds of copies of the repeat) are required to produce type 2 diseases. The underlying mechanism of these diseases seems to be altered gene expression, and it is likely to occur at different stages for each disorder.

Although the mutation events can be characterized as those that undergo relatively short expansions (type 1) compared to those capable of very large expansions (type 2), type 2 disease alleles do exhibit short increases in length. An attractive hypothesis is that all triplet repeats can undergo short expansions during meiosis, and postmeiotic mechanisms produce the large expansions observed at type 2 loci (Ashley and Warren 1995; Paulson and Fischbeck 1996); possibly large expansions cannot occur for type 1 loci due to constraint produced by the translation of the CAG codon into a polypeptide.

The genetic instability of a specific TRS is dependent on its locus, and expansions of TRS are not due to a destabilization of the whole genome (Loeb 1994; Sutherland and Richards 1995). Therefore, it is apparent that the genetic instabilities associated with triplet repeat diseases do not reflect those observed at microsatellite sequences in general (see Sect. 1.2). Clearly, there are many factors controlling the genetic stability of trinucleotide repeats, but a major determinant is the length of the repeat sequence. The majority of the population have alleles that are genetically stable, while disease families have longer alleles that are unstable. The existence of a critical threshold for disease formation is substantiated by studies which show that interruptions within the repeat sequence stabilize its genetic propagation (Chung et al. 1993; Eichler et al. 1994; Kunst and Warren 1994). The correlation between increasing repeat size and disease severity is best illustrated by the type 1 disorders, although analysis of the spinocerebellar ataxias shows that each disease has specific characteristics (see above). Type 2 disorders have a wider range of observed repeat lengths, but the most severe diseases are still associated with the largest number of repeats. The existence of similar threshold sizes for the various triplet repeat diseases supports the notion that common mechanisms produce repeat tract instability in each disorder.

Examination of different tissues and cell lines has shown variations in the genetic instability of their TRS. All tissues examined for the HD repeat displayed repeat mosaicism, with the greatest instability in brain and sperm (Duyao et al. 1993; Telenius et al. 1994). Single cells from male myotonic dystrophy patients were found to have various sizes of TRS, with a directional bias towards increasing length in somatic tissues (Monckton et al. 1995). Analysis of transgenic mouse lines with large repeats derived from the human myotonic dystrophy gene revealed a high degree of genetic instability in germline and somatic cells (Monckton et al. 1997). Interestingly, germline variation of the myotonic dystrophy TRS included frequent deletions to the repeat (Monckton et al. 1995, 1997). These observations of germline and somatic mosaicism within particular TRS suggest that cell-specific factors influence the genetic stability of these repeats.

Much effort has been devoted to the development of animal models of triplet repeat diseases in the hope that these will help us understand their molecular mechanisms. Inactivation of the gene associated with fragile X syndrome in humans (FMR1) caused mice to display abnormalities similar to those of human patients (Dutch-Belgian Fragile X Consortium 1994) and thus may prove valuable in determining the physiological pathways involved in fragile X syndrome. Disruptions to the murine homologue of the HD gene showed that it was essential for normal embryonic development (Duyao et al. 1995; Nasir et al. 1995; Zeitlin et al. 1995), and mice that transcribed a human HD cDNA with 44 CTG·CAG repeats did not develop HD symptoms (Goldberg et al. 1996). Assuming that the mouse is an appropriate animal model in which to study HD, the lack of an HD phenotype in these studies suggests that the disorder is due to an altered interaction of the HD protein and that this may relate to a secondary function.

A number of studies have generated mice that are transgenic for proteins carrying expansions of polyglutamine tracts. The absence of phenotype in mice carrying the normal or expanded protein associated with SBMA may have been due to a nonphysiological pattern of expression (Bingham et al. 1995). A neurological phenotype was observed in mice transgenic for SCA1 (Burright et al. 1995) and SCA3 (MJD; Ikeda et al. 1996). Recently, mice that were transgenic for the 5' end of the human HD gene have been shown to exhibit many of the features of HD (Mangiarini et al. 1996). However, in transgenic mice studies in which repeat length was examined, the TRS were stable upon genetic transmission (Bingham et al. 1995; Burright et al. 1995; Goldberg et al. 1996; Ikeda et al. 1996). These studies provide further evidence that the relationship between repeat instability and disease formation is complex and is likely to involve the interaction of numerous cellular factors.

1.2
Microsatellite Instability and Cancer

All organisms have biochemical systems that provide a high fidelity of genome replication and thus prevent the propogation of mutated phenotypes (Umar and Kunkel 1996). Mechanisms are in operation to ensure high fidelity during DNA synthesis and also to correct mistakes that are made during replication. Deficiencies within these systems have the potential to produce many errors within genome sequences, and this situation has been termed the "mutator phenotype," since it would produce an organism prone to mutations. The progression of many human tumors through multiple stages (Vogelstein and Kinzler 1993) suggests that their development requires a number of genetic alterations. The existence of a mutator phenotype would increase the potential for tumors to arise and therefore may be a primary event in tumor occurrence (Loeb 1994).

Length changes are frequently observed within simple repeating sequences in all genomes. However, elevated frequencies of length changes occur under some conditions of reduced replication fidelity. A fundamental system involved in maintaining genomic integrity is that first identified in E. coli as methyl-directed mismatch repair (Modrich 1991). A number of experiments have shown a similar mismatch repair system in eukaryotes, with a high degree of conservation throughout all organisms (Fishel and Kolodner 1995; Kolodner 1995; Modrich and Lahue 1996; Umar and Kunkel 1996). Upon inactivation of this system of DNA repair, increased heterogeneities have been observed at some unit lengths of simple repetitive DNA (e.g., mono- and dinucleotides) in bacterial systems (Levinson and Gutman 1987a; Freund et al. 1989) and in yeast (Strand et al. 1993, 1995). Since there is an increased rate of mutation throughout the whole genome under these conditions, these observations suggest that the level of heterogeneity at simple repeating sequences is a good indicator of the overall level of DNA stability in cells.

Experimental evidence that deficient DNA repair could cause human tumors was obtained when cell lines from inherited and sporadic human cancers were shown to undergo an increased frequency of length changes in specific dinucleotide repeats (Aaltonen et al. 1993; Ionov et al. 1993; Thibodeau et al. 1993). It was later shown that this increased genetic instability was due to defects in mismatch repair proteins (Fishel et al. 1993; Leach et al. 1993; Parsons et al. 1993). Numerous studies have confirmed that defective mismatch repair is responsible for the elevated microsatellite instability and increased mutation rate in some cancers. For example, mice containing homozygous null mutations in some mismatch repair genes are viable, but develop tumors and exhibit elevated microsatellite instability (Baker et al. 1995; de Wind et al. 1995; Reitmair et al. 1995).

The associations of defective mismatch repair and elevated microsatellite instability are particularly strong for hereditary nonpolyposis cancer, one of the most common inherited disorders known (de la Chapelle and Peltomaki 1995; Marra and Boland 1995). A number of recent reviews have discussed the relationship between microsatellite instability and cancer formation (Eshelmann and Markowitz 1996; Kinzler and Vogelstein 1996). Presently, it is not clear whether elevated microsatellite instability can be used as an absolute marker for defective mismatch repair. Although some cancers with elevated microsatellite instability do not carry a defect in the known mismatch repair genes, only a subset of genes involved in human mismatch repair have been identified (Kolodner 1995; Eshelmann and Markowitz 1996). Moreover, the effects of other DNA repair systems on microsatellite instability in humans is unknown, and it is conceivable that these will play a role in tumor development.

Genetic instabilities within mono- and dinucleotide repeats increase for longer runs of consecutive repeats and are decreased by interruptions to the repeat sequence (Levinson and Gutman 1987b; Umar and Kunkel 1996). These observations are consistent with the hypothesis that slipped-strand mispairing during DNA synthesis generates misaligned intermediates (see above). These parameters are intrinsic to the DNA repeat, but it is also known that flanking sequences can influence the genetic stability of simple repeat sequences (Umar and Kunkel 1996). As discussed below in relation to TRS, these observations suggest that many factors, including DNA repair, replication, and transcription, affect the genetic stability of microsatellite sequences.

2
Instabilities in a Genetically Defined Bacterial System

2.1
General Comments on CTG·CAG

As described above, progress has been made in our understanding of genetic instabilities from investigations with lower eukaryotes (yeast), cultured mammalian cells, transgenic mice models, and analyses of sperm. The optimum system would be a genetically controllable higher eukaryotic system. However, almost by definition, this is not available at the present time. Thus maximum information may be derived at present by investigations on simpler bacterial systems for evaluating the molecular processes that elicit expansions, whereas the more complex eukaryotic analyses will provide useful information on the timing of events during development, the sex of parents determination of expansion, germline and somatic hypermutability, genome position effects, etc.

This laboratory has focused attention on investigations of molecular instabilities of CTG·CAG and CCG·CGG repeat sequences in a genetically and biochemically defined organism (*E. coli*; reviewed in Wells 1996). Kang et al. (1995a) established a defined genetic system which shows promise for the molecular dissection of this process. To study expansions, these workers determined whether a plasmid that contains $(CTG·CAG)_{130}$ is completely homogeneous as a cloned molecule or whether deletions and expansions had occurred that gave rise to sequence heterogeneity, even in a tiny percentage of the molecules. The insert containing the triplet repeat was excised from the vector and separated by gel electrophoresis. The regions of the gel either above or below the insert band were eluted and "recloned"; recombinant plasmids were obtained that contained successively larger or smaller inserts, respectively. The family of inserts characterized by these methods contained repeat units ranging from 17 to 300. Hence, expansion and deletion occur in *E. coli*. This discovery lays the foundation for evaluating host cell genetic factors (e.g., replication, recombination, mismatch repair) that may elicit genetic instabilities.

The frequency of genetic expansions or deletions in *E. coli* depends on the direction of replication. Large expansions occur predominantly when the CTGs are in the leading template strand rather than the lagging strand. However, deletions are more prominent when the CTGs are in the opposite orientation. It should be noted that both deletions and expansions occur in both orientations. Most deletions generate products of defined size classes. Strand slippage coupled with nonclassical DNA structures probably accounts for these observations and relates to expansion/deletion mechanisms in eukaryotic chromosomes. DNA sequence analyses showed that expansion and contraction always occurred in multiple repeats of 3 bp. Prior investigations (Jaworski et al. 1991) showed that deletions in dinucleotide repeat sequences occurred in multiple units of 2 bp.

A possible mechanism for the expansion and deletion behaviors has been proposed (Fig. 1; Kang et al. 1995a). For expansion, a hairpin loop may form on the lagging strand nascent DNA (CTG strand). NMR investigations (Smith et al. 1995) revealed that CTG oligomers form a stable antiparallel duplex with TT pairs, whereas the complementary CAG strand forms a metastable conformation. When CTG is the lagging strand template (orientation II), a loop may form on the lagging strand which will be bypassed during DNA synthesis to generate deletions. Multiple slippages (Wells and Sinden 1993) may be promoted by an "idling polymerase" caused by a strong block such as a DNA structure or the presence of proteins, which causes continuous slippage (primer realignment), resulting in the expansion of larger sequences. Recent work in yeast with similar TRS has verified this model (Freudenreich et al. 1997). Other workers (Jeffreys et al. 1994) suggested that gene conversion events explain germline mutations at human minisatellites; however, these two mechanisms are not mutually exclusive.

Recent investigations (Bowater et al. 1996) revealed the relationship between cell growth and deletions of CTG·CAG triplet repeats in plasmids. Long CTG·CAG repeats in plasmids can influence cell growth, which results in the observed expansions and deletions. During extended growth periods, the observed frequencies of deletion were dramatically increased if the cells passed through stationary phase before subculturing. High frequencies of deletions were observed because of a growth advantage of cells containing plasmids with deleted triplet repeats. These observations (Bowater et al. 1996) are the first to show a direct influence between a plasmid based DNA sequence or structure and factors controlling bacterial growth. Additional studies (Bowater et al. 1997) showed that transcription promotes deletions of long CTG·CAG triplet repeats from human neuromuscular disease genes. However, a lower frequency of deletions is found also in the absence of transcription. These investigations were performed in recombinant plasmids that contained inducible promoters, and the frequency of deletions was monitored biochemically. These studies suggest a role for the involvement of transcription in DNA strand slippage which gives rise to deletions and expansions.

Furthermore, single-stranded DNA-binding protein enhances the stability of CTG·CAG triplet repeats in *E. coli* (Rosche et al. 1996). Studies were conducted with mutants that lack SSB protein in order to evaluate the possible involvement of this protein, which is an important component in DNA replication, repair, and recombination. SSB can prevent the formation of DNA secondary structures. Replication can pause at sites of potential DNA secondary structure, and pause sites are associated with template misalignment mutagenesis. The potential for slippage and for DNA secondary structure formation within TRS may contribute to the instabilities of these sequences observed in individuals with triplet repeat diseases (described above). With a biochemical assay for stability, Rosche et al. (1996) showed that the absence of single-stranded DNA-binding protein leads to an increase in the frequency of large deletions within the TRS.

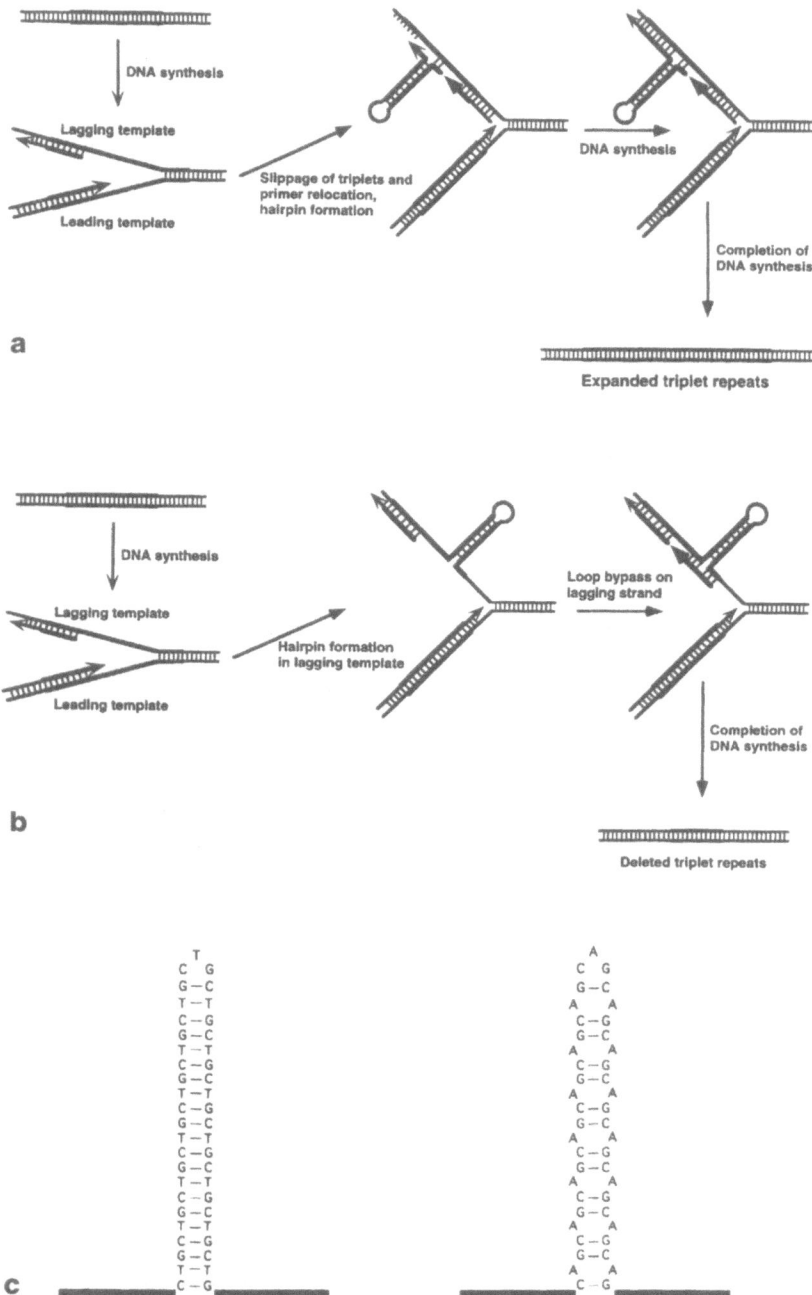

Fig. 1a–c. A model for orientation-dependent instability of CTG repeats during replication. **a** Expansion (orientation I). **b** Deletion (orientation II). **c** Hairpin structure. *Left*, More stable; *right*, less stable. (Wells 1996, with permission)

2.2
Preferential Expansion of CTG·CAG

Ohshima et al. (1996a) discovered that the CTG·CAG triplet repeat is the dominant genetic expansion product. This extraordinary discovery was made possible by the successful cloning and characterization of all ten TRS (Ohshima et al. 1996b, c). The relative capacity of the ten TRS to be expanded in *E. coli* (Kang et al. 1995a) was explored with a competition study. Surprisingly, the CTG·CAG triplet repeat was expanded at least nine times more frequently than any of the other nine triplets (Ohshima et al. 1996a). Low levels of expansion were found also for GTG·CAC, GTC·GAC, CGG·CCG, and GAA·TTC. Thus the structure of the CTG·CAG repeat and/or its utilization by the DNA synthetic systems in vivo must be quite different from the other triplets. The surprising discovery that CTG·CAG triplet repeats are the dominant expansion products, as found in clinical samples from human hereditary diseases (reviewed in Wells 1996), suggests the importance of DNA structural properties (Wells 1988, 1996; Sinden 1994;). Other investigations have revealed that duplex CTG·CAG and CGG·CCG repeats have unorthodox properties, including nucleosome assembly (Wang et al. 1994, 1996), the capacity to cause DNA polymerases to pause within the repeat sequences (Kang et al. 1995b; Ohshima and Wells 1997), and conformational features as revealed by helical repeat and polyacrylamide gel migrations (Chastain et al. 1995; Bacolla et al. 1997). Further elucidation of the CTG·CAG repeat structural features along with the genetic factors responsible for expansion may explain why most triplet repeat hereditary disease genes contain CTG·CAG repeats (reviewed in Wells 1996). Although other triplet repeats are found in the human genome (Gastier et al. 1995), the lengths are shorter (generally fewer than 15 repeats) than found for these disease genes.

   Both CTG·CAG and CGG·CCG have topological properties of writhe and flexibility that have not been described for other DNA (see Sect. 3). In addition, the former TRS is unusual in its high binding affinity for nucleosomes and is stabilized by mismatch repair-deficient cells. The latter two properties are not shared with CGG·CCG. Future work will be required to evaluate the role of these behaviors in the preferential expansion of CTG·CAG.

2.3
Site of Expansion

Kang et al. (1996) described an investigation aimed at identifying the region of CTG·CAG triplet repeats that is preferentially expanded. Interestingly, the repeats are expanded distal to the replication origin (ColE1) as a single large event. Analysis of expanded regions using the interrupting CTA triplet sequence as a location marker within the CTG·CAG tract revealed that the expansion of large CTG·CAG repeats is one event rather than an accumulation of multiple small expansions and that the expansions occur more frequently in the region distal from the replication origin. In addition, we showed that a loss of interruptions increases the expansion frequency. Thus the instability of large triplet repeats in hereditary diseases occurs by a mechanism different from the instability in microsatellite sequences caused by defects in mismatch repair systems for certain sporadic cancers and hereditary nonpolyposis colorectal cancers.

2.4
Mismatch Repair

Mismatch repair deficient *E. coli* (Modrich and Lahue 1996) were studied in order to further elucidate the factors involved in genetic instabilities as well as DNA structural issues in vivo (Jaworski et al. 1995). Long CTG·CAG repeats are stabilized in ColE1-derived plasmids in *E. coli* containing mutations in the methyl-directed mismatch repair genes (*mutS*, *mutL*, or *mutH*). When plasmids containing (CTG·CAG)$_{180}$ were grown for about 100 generations in *mutS*, *mutL*, or *mutH* strains, 60–85 % of the plasmids contained a full-length repeat, whereas in the parent strain only about 20 % of the plasmids contained the full-length repeat. The deletions occur only in the (CTG·CAG)$_{180}$ insert, and not in DNA flanking the repeat. While many products of the deletions are heterogeneous in length, preferential deletion products of about 140, 100, 60, and 20 repeats were observed. The *E. coli* mismatch repair proteins apparently recognize three-base loops formed during replication and then generate long single-stranded gaps where stable hairpin structures may form; these can be bypassed by DNA polymerase during the resynthesis of duplex DNA (Fig. 2). Direct experiments will be required to test the veracity of this model. Similar studies were conducted with plasmids containing CGG·CCG repeats; no stabilization of these triplets was found in the mismatch repair mutants. The reason for this was unclear, but may be due to the rate of formation and the stability of hairpins in CTG·CAG and CGG·CCG repeats (designated by question marks in Fig. 2). Since prokaryotic and human mismatch repair proteins are similar (Modrich and Lahue 1996), and since several carcinoma cell lines which are defective in mismatch repair show instability of simple DNA microsatellites (Jaworski et al. 1995; Modrich and Lahue 1996), these mechanistic investigations in a bacterial cell may provide insights into the molecular basis for some human genetic diseases.

2.5
Fragile X CGG·CCG

A series of inserts containing six to 240 copies of CGG·CCG were stably cloned in plasmids. Several factors influence the stability (deletions and expansions) of the inserts; repeat length, the presence of interruptions, the orientation of the insert relative to the unidirectional replication origin, *E. coli* host strains, the location of the insert, and the copy number of the vector. The instability varies strongly with the length of the insert; longer tracts of CGG·CCG repeats show a greater degree of instability compared to shorter inserts. Furthermore, the effect of the length of DNA polymerase pausing was also observed during synthesis of the repeat in vitro when the Klenow fragment of DNA polymerase I

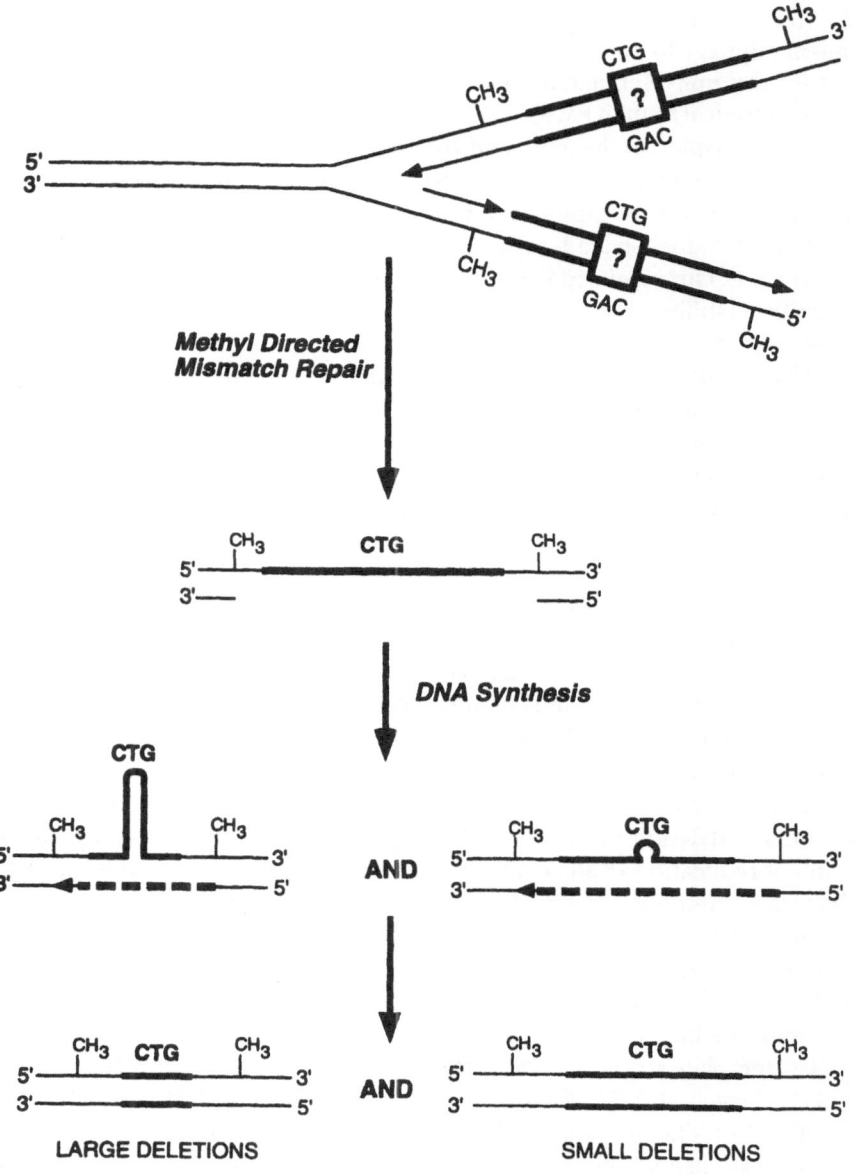

Fig. 2. Models for the involvement of *E. coli* methyl directed mismatch repair proteins in the enhancement of destabilization of (CTG)$_n$ triplet repeats in vivo. (Jaworski et al. 1995, with permission)

was used; lengths of greater than 61 repeats showed stronger pausing sites, occurring at repeat number 30 (away from the CGG·CCG start site), when CCG was the template strand. This phenomenon was also observed with CTG·CAG triplet repeats (Kang et al. 1995b). These results suggest that, at a critical length, the CGG sequence adopts a non-B conformation or conformations (see Sect. 2.6) which block DNA polymerase progression, leading to the idling and subsequent slippage to give expanded products and hence provide the molecular basis for this non-Mendelian genetic process.

The canonical human FMR-1 repeat carries 30 CGG·CCG triplets interrupted by two AGG triplets at the tenth and 20th repeat. Fragile X carriers carry longer repeats (50–200) containing long stretches of uninterrupted CGG·CCG triplets which predispose this sequence to hyperexpansion in successive generations. Affected individuals have longer methylated repeats (230–2000; Warren and Nelson 1994). Our results indicate that the presence of interruptions greatly enhances the stability of the CGG·CCG tract in *E. coli*. Other studies on the alleles derived from human patients show the presence of stable and unstable CGG·CCG triplets of similar size, suggesting that a feature other than length, but intrinsic to the repeat, was responsible for stability. This supported the observations made by Eichler et al. (1995), who found that lengths of more than 33 uninterrupted CGG·CCG triplets showed marked instability, regardless of total repeat length, suggesting that loss of the AGG interruptions is an important mutational event in the generation of alleles predisposed to the fragile X syndrome.

As mentioned above, another important factor dictating stability is the orientation of the CGG·CCG-containing insert. Our results indicated that the triplet repeat was stably maintained in vectors if the CGG strand was in the leading template strand (orientation I) with respect to the origin of replication. However, if CCG fell in the leading template strand (orientation II), the insert was highly destabilized (depending on the length), undergoing deletions and expansions. As in the case of CTG·CAG-repeating sequences, the frequency of expansion and deletion of the CGG·CCG triplet repeats is influenced by the direction of replication (Kang et al. 1995a), which involves an asymmetric DNA polymerase complex that simultaneously replicates both the leading and the lagging strand (Wells and Sinden 1993). Replication-dependent deletion between direct repeats occurs preferentially in the lagging strand due to the unequal probability of forming hairpins (Trinh and Sinden 1991). Therefore, the deletion of the insert (in orientation II) can be explained by the propensity of the CGG template strand to form a stable hairpin (Fry and Loeb 1994; Chen et al. 1995; Gacy et al. 1995; Mitas et al. 1995; Mitchell et al. 1995;) which is bypassed by the replication machinery during resynthesis of the DNA. On the other hand, expansions within the tract are likely due to strand realignment through slippage of the complementary strands during pausing (described above) to generate a folded and elongated nascent DNA on the leading strand

(Kang et al. 1995a). Deletions were the most abundant species detected, but expansions were also visible when pRW3024, i.e., (CGG·CCG)$_{24}$ in orientation II, was propagated in *E. coli* DH5α; the bands differed from each other by one repeating CGG·CCG unit, suggesting the involvement of slipped structures during replication. This method allowed the cloning of the expanded and deleted products (6 to 49 repeats) in orientation I and their propagation in *E. coli* SURE to give a stable DNA preparation.

2.6
DNA Polymerase Pausing

The pausing of DNA synthesis in vitro at specific loci in double stranded CTG·CAG and CGG·CCG triplet repeats was found serendipitously (Kang et al. 1995b). The DNA syntheses of CTG·CAG triplets ranging from 17 to 180 and CGG·CCG repeats ranging from 9 to 160 repeats in length were studied in vitro. Primer extensions using the Klenow fragment of DNA polymerase β, the modified T7 DNA polymerase (Sequenase), or the human DNA polymerase (paused strongly at specific loci in the CTG·CAG repeats. The pausings were abolished by heating at 70 °C. As the length of the triplet repeats in duplex DNA, but not in single-stranded DNA, was increased, the magnitude of pausings increased. CGG·CCG triplet repeats also showed similar, but not identical patterns of pausings. These results indicate that appropriate lengths of the triplets adopt a non-B conformation or conformations that block DNA polymerase progression; the resultant idling polymerase may catalyze slippages to give expanded sequences and hence provide the molecular basis for this non-mendelian genetic process. In addition, in vivo replication studies in *E. coli* (S.M. Mirkin, pers. comm.) with plasmids containing the CGG·CCG repeat revealed length-dependent pause sites; the nature of the DNA conformation which generates these pauses in vivo has not been characterized. Additional recent in vitro investigations (Ohshima and Wells 1997) proved that the product of pausing was a hairpin structure caused by primer realignment, loop formation with reannealing of the nascent DNA to itself, and synthesis on the newly formed strand. In summary, the in vitro replication behavior of TRS is fascinating; our recent work with the Friedreich's ataxia GAA·TTC sequence shows the pausing behavior, presumably due to triplex formation (Ohshima et al. 1996b).

Ohshima and Wells et al. (1997) have isolated and analyzed the products of paused synthesis found at approximately 30–40 triplets from the beginning of the TRS. DNA sequence analyses revealed that the paused products contained short tracts of homogeneous TRS (6–12 repeats) in the middle of the sequence corresponding to the flanking region of the template-primer system. The sequence at the 3' side terminated at the end of the primer, indicating that the primer molecule had served as template. In addition, chemical probe and polyacrylamide gel electrophoretic analyses revealed that the paused products existed in hairpin structures. We postulate (Fig. 3) that paused products for the

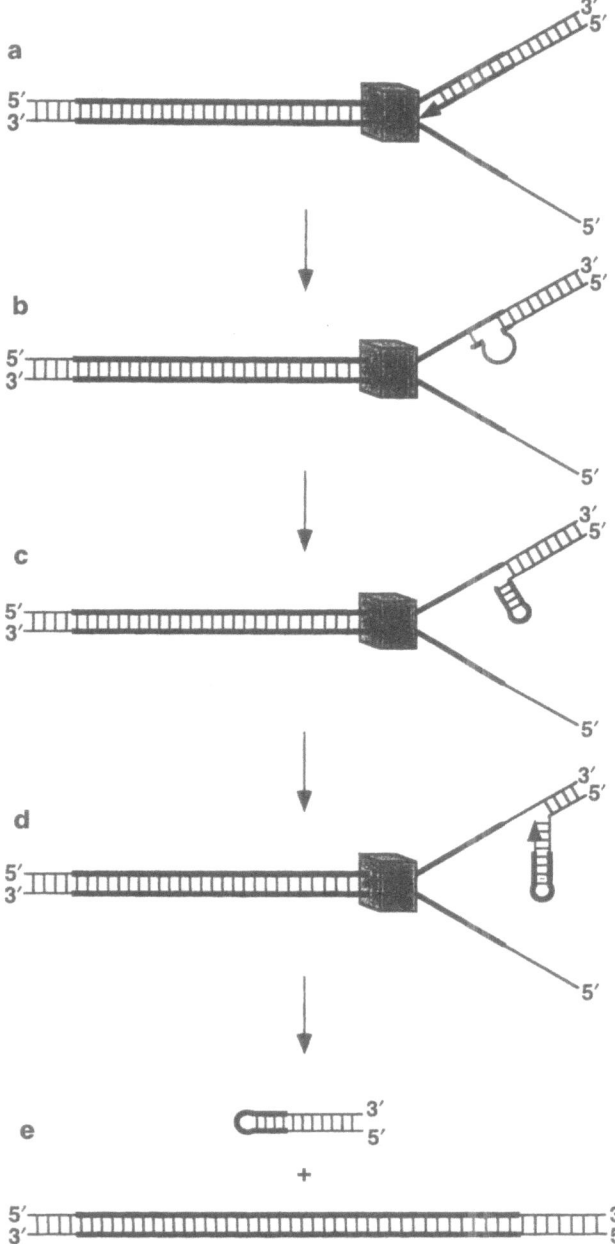

Fig. 3a–e. Model for formation of hairpin structures at sites of paused DNA synthesis followed by strand slippage and primer realignment. **a** Pausing of DNA synthesis. **b** Dissociation of the nascent strand of triplet repeat sequences by slippage. **c** Hairpin formation and primer realignment. **d** DNA chain elongation. **e** Termination of DNA synthesis. (Ohshima and Wells 1997, with permission)

DNA polymerases are caused by the existence of an unusual DNA conformation or conformations within the TRS (see Sect. 3) during in vitro DNA synthesis, enhancing the DNA slippages and the hairpin formation in the TRS due to primer realignment. The consequence of these steps is DNA synthesis to the end of the primer and termination. Primer realignment, including hairpin formation, may play an important intermediate role in the replication of TRS in vivo to elicit genetic expansions.

A logical reason for the impediment in long CTG·CAG and CGG·CCG sequences is flexible and writhed DNA (Bacolla et al. 1997; Gellibolian et al. 1997; see Sec. 3). However, we have not directly proven that this structural feature is responsible for the DNA polymerase pausing.

2.7
Molecular Similarities Between Humans and Escherichia coli

The studies described above on a genetically and biochemically tractable system for elucidating the molecular mechanisms responsible for expansion, and thus anticipation, represent a significant advance. Several remarkable molecular similarities exist, including the following (reviewed in Wells 1997):

- TRS (CTG·CAG, CGG·CCG, or AAG·CTT) are genetically unstable (expansions and deletions).
- Longer repeats are more unstable than shorter sequences.
- CTG·CAG is preferentially expanded in *E. coli*; this repeat sequence was found in six of the nine triplet repeat diseases.
- Repeat sequence imperfections (polymorphisms) stabilize long tracts of TRS.
- Similar types of imperfections (polymorphisms) are found (e.g., the
- polypurine·polypyrimidine motif is maintained in the Friedreich's ataxia AAG·CTT repeat sequence).
- The lengths of the smallest deletion products in *E. coli* (ten to 20 triplet repeats) approximate the lengths found in normal humans.
- DNA polymerases from humans and *E. coli* pause in long CTG·CAG, CGG·CCG, and AAG·CTT sequences, thus rendering them susceptible to mutations.

Hence, certain features of the molecular processes related to the involvement of TRS in human hereditary diseases may be elucidated effectively in simple cellular systems. Obviously, a number of other developmental and neurobiological questions can only be solved in higher eukaryotic cells. Thus some features of the concept of "unstable genes, unstable mind" may be tractable in genetically defined systems in mice, microbes, and molecules.

2.8
Summary of Factors Influencing Genetic Instability

As elaborated above, several factors are known to influence the stability of long TRS in plasmids in *E. coli*. These factors include the following:

- *Genetic makeup of host cells*. The absence of recA is vital for the cloning of long TRS and, as discussed above, the presence of SSB protein is significant (Rosche et al. 1996).
- *Growth conditions*. Several factors related to growth conditions including media and not permitting the cells to go through stationary phase, are also critical (Bowater et al. 1996, 1997).
- *Generations of cells*. Since the deletion and expansion behavior appears to be due to DNA replication, the number of generations of cells (Kang et al. 1995a) is important.
- *Transcription*. Active transcription through the TRS appears to enhance deletions (Bowater et al. 1997).
- *MutS, MutH, and MutL*. The absence of these mismatch repair functions enhances stability, since it is likely that short loops which are formed by DNA slippage are recognized and cleaved with deletions as the products.
- *Nucleotide excision repair*. It is possible that nucleotide excision repair may also be involved in the observed instabilities (P. Parniewski and R.D. Wells, unpubl.).
- *Vector and cloning location*. The type of vector and the location of cloning of a TRS in the vector is important and may relate to copy number. The instabilities may be due, in part, to the proximity to DNA polymerase I-III switch sites (Jaworski et al. 1995).
- *Sequence of insert*. Certain types of sequences are more unstable than others. For example, the fragile X CGG·CCG sequence appears to be quite unstable, whereas several other sequences are somewhat more stable in the form of long tracts (Ohshima et al. 1996b, c; Shimizu et al. 1996).
- *Length of insert*. The length of the insert is critical; shorter lengths (30–50 TRS) are rather more stable, but longer sequences (>200 TRS) are quite unstable.
- *Interruptions in insert*. The presence of interruptions, or polymorphisms as recognized in human genetics, has a stabilizing effect in long TRS. This behavior has been observed with virtually all of the ten TRS studied to date (Shimizu et al. 1996; Ohshima et al. 1996b, c).
- *Orientation of insert*. The orientation of the insert is important for favoring deletions or expansions, as described above.

3
Flexible and Writhed CTG·CAG and CGG·CCG

Whereas the results described above implicate DNA structural features in the process of destabilization of the TRS, analyses conducted with chemical, enzymatic, and structural probes that detect perturbed or single-stranded regions failed to demonstrate that CGG·CCG and CTG·CAG adopt any of the non-B DNA structures identified so far, which include left-handed Z-DNA, inter and intramolecular triplexes, cruciforms, tetraplexes, and nodule DNA (Wells 1988, 1996; Bacolla et al. 1997). On the other hand, linear fragments containing sufficiently long CTG·CAG (and to a lesser extent CGG·CCG) migrated faster than expected on polyacrylamide gels (Chastain at al. 1995; Bacolla et al. 1997), indicating that these sequences possess peculiar structural properties. Such properties were elucidated from experiments on circularization kinetics (CK) and apparent helical repeat determination.

CK measures the rate of ring closure of linear DNA fragments that have complementary, single-stranded ends and enables the flexibility of DNA to be determined, whereas the determination of the apparent helical repeat gives a qualitative indication of writhe, i.e., the ability of the DNA helix to deflect from planarity in three-dimensional space. Before describing these results, we will outline the relevance of DNA flexibility in general biological processes.

3.1
Flexibility of DNA and General Biological Processes

The flexibility of DNA is essential for the organization of genetic information. Linear duplex DNA is assembled into high-order chromatin structures in both prokaryotic and eukaryotic cells, which enables selective (e.g., tissue- and development-specific) gene expression. In addition, a high level of condensation is achieved in the eukaryotic nuclei during segregation of metaphase chromosomes. Here, a 2.5-μm-long chromatid contains a single copy of a DNA molecule about 30 mm in length when fully extended (DuPraw 1974). The required compaction factor, from 30 mm to 2.5-μm, of 12000-fold obviously involves dramatic bending of the DNA molecule.

Binding between protein complexes distally located along the DNA provides a means for transcriptional control. A clear example is the regulation of the arabinose operon in *E. coli*. An interaction is necessary between two araC proteins, one bound to *araI*, located just upstream from the initiation of transcription, and the other to *araO_2*, situated 211 bp further upstream. In the absence of arabinose, araC bound to *araI* and *araO_2*, which stably dimerize as a result of DNA looping; this complex leads to repression of transcription. The presence of arabinose weakens the protein contacts, disfavors DNA looping, and results in high levels of transcription (Schleif 1992). Similar looped structures are also thought to constitute a mechanism by which enhancer-mediated stimulations of transcription operate (Schleif 1992).

Initiation of DNA replication necessitates unwinding of the double helix. Studies in viruses, bacterial cells and yeast cells all reveal that initiation of replication is contingent upon the presence of a DNA unwinding element (DUE; DePamphilis 1993). The function of DUE is to provide a sequence in which the separation (melting) of the two strands is thermodynamically favorable. A region of nonhelical, single-stranded DNA is required for the onset and initial progression of a replication fork (DePamphilis 1993). This melting operation involves unwinding (untwisting) of the helix, which is accomplished by one or two cooperative activities: DNA helicase and negative supercoiling. DNA helicases physically separate the two DNA strands, whereas negative supercoiling (which is essential in certain systems; Alfano and McMacken 1988) lowers the activation energy of the enzymatic process.

Thus supercoiling is an important feature of DNA. Regardless of whether the linear or circular chromosomal DNA is involved, rotation of the strands about a free end (such as a nick or the chromosomal ends) is much restricted in vivo. It follows that unwinding activities, such as the one on DUE, force adjacent segments of the DNA to overwind, which causes the duplex DNA to supercoil. Due to a continuous requirement for unwinding, processes such as replication and transcription are associated with the constant accumulation of supercoils in front of the respective enzymatic complexes, which may prevent further enzyme translocation. These supercoils are normally removed by topoisomerases (Wang 1996). Studies on topoisomerases have led to the appreciation that supercoiling plays both a favorable and an unfavorable role in general cellular functions, including replication, transcription, recombination, excision repair, chromatin organization, and genome stability (Wang 1996).

3.2
Determination of DNA Flexibility

The determination of flexibility is a problem of physical chemistry and entails evaluating the forces that oppose bending and twisting of the double helix. Four techniques have provided estimates of such forces (moduli) for B-DNA of random composition: (1) transient electric birefringence and dichroism (TEB/TED), (2) electron microscopy (EM), (3) CK, and (4) Monte Carlo simulation (MC), a computational approach (reviewed by Diekmann et al. 1982; Hagerman 1988; Hagerman and Ramadevi 1990; Taylor and Hagerman 1990; Kahn et al. 1994; Bednar et al. 1995; Hodges-Garcia and Hagerman 1995). These studies monitor the behavior of short (few hundred bp long) linear DNA fragments under different conditions and quantitate the moduli by applying theoretical treatments to experimentally measured parameters. Of these, CK has become increasingly popular due to its simple execution (Shore et al. 1981; Shore and Baldwin 1983a) and to the development of comprehensive analytical solutions to the underlying statistical mechanical theory (Shimada and Yamakawa 1984, 1985). CK measures the fraction of linear molecules that are converted into circles as a function of time by the action of T4 DNA ligase on complementary single-stranded ends. We have employed this method to measure the bending and twisting forces of DNA composed primarily of $(CTG \cdot CAG)_n$ and $(CGG \cdot CCG)_n$ repeats (Bacolla et al. 1997).

3.3
Bending and Twisting Forces of (CTG·CAG)$_n$ and (CGG·CCG)$_n$

3.3.1
Theory of Ring Closure

DNA is in constant motion in solution due to thermal energy. At any time, the end of a linear molecule may either encounter the end of a different molecule (bimolecular encounter) or its other end (cyclization encounter). *Bimolecular encounter* depends on temperature (higher temperature favors more encounters) and the total concentration of ends (more ends, more encounters), whereas *cyclization encounter* is dependent on temperature, but is independent of end concentration. In kinetic terms, the former process is second order, whereas the latter is first order (Jacobson and Stockmayer 1950; Wang and Davidson 1966). Additionally, cyclization encounter is affected by the length of the molecule, its stiffness, and the fractional helical turn. These factors are considered separately.

Length. The two ends of a 10 bp-long DNA molecule will not circularize; it is like trying to bend a 1-cm-long nail into a circle. In a 1000 bp-long molecule, the chances are that the two ends will meet sooner or later; analogously, we can bend into a circle a 10 m-long rod made of the same material as our previous 1 cm-long nail. Therefore, very short molecules do not cyclize efficiently, whereas longer molecules do. However, the consideration that "the longer the molecule, the more frequent the cyclization" is only valid within a fixed range of length. In fact, if we keep one end of the molecule fixed at one position and follow the trajectory of the other end, we would see it traveling through a space volume whose outer border is determined by the extended length of the molecule. If we consider the probability that the traveling end will come in contact with the fixed end, two effects counteract each other. The longer the molecule, the more movement is allowed, which increases cyclization. However, the longer the molecule, the greater the volume "explored" by the free end, which reduces cyclization. This problem, which concerns the behavior of any elastic rod, has been studied both theoretically and computationally (Shimada and Yamakawa 1984). Curve A in Fig. 4 shows how the probability of encounter of two ends of DNA varies with the length of the DNA. At short lengths, the probability increases with length, but beyond a certain limit it decreases. This decrease is due to the "takeover" effect by the sphere volume.

Stiffness. Stiffness increases the average end-to-end distance of an elastic rod and therefore decreases the probability of cyclization. Curve B in Fig. 4 shows how this probability varies for molecules that are twice as stiff as A. Whereas the short members do indeed circularize less efficiently than A, the sphere volume effect becomes dominant at longer lengths, to the point of abolishing the differences between A and B. This phenomenon is important and shows that, in order to study the flexibility of DNA, the length of the fragments used for analyses should not extend far beyond the peak of circularization probability.

Fractional Helical Turn. Consider two DNA molecules, one with an integral number of helical turns and the other 5 bp longer. During cyclization encounters, the free end of the first molecule is properly oriented for circle formation with its fixed end, whereas the free end of the second molecule is now turned 180 ° away from the fixed end. Thus, in contrast to the first molecule, circularization of the second molecule requires that the free end both collides with and rotates relative to the fixed end. Curve C in Fig. 4 shows the probability of ring closure for molecules as stiff as A, but whose free ends need to rotate by 180°. The differences between A and C are remarkable at short lengths, the values being much lower for C. Thus both the bending and the twisting forces may be evaluated by CK. Once again, the differences vanish at long lengths, further stressing the remarks pointed out previously. Since with real DNA molecules the fractional helical turn oscillates from 0 to 0.5 with length, the resulting probability follows an oscillatory pathway between an upper (A) and a lower (C) boundary.

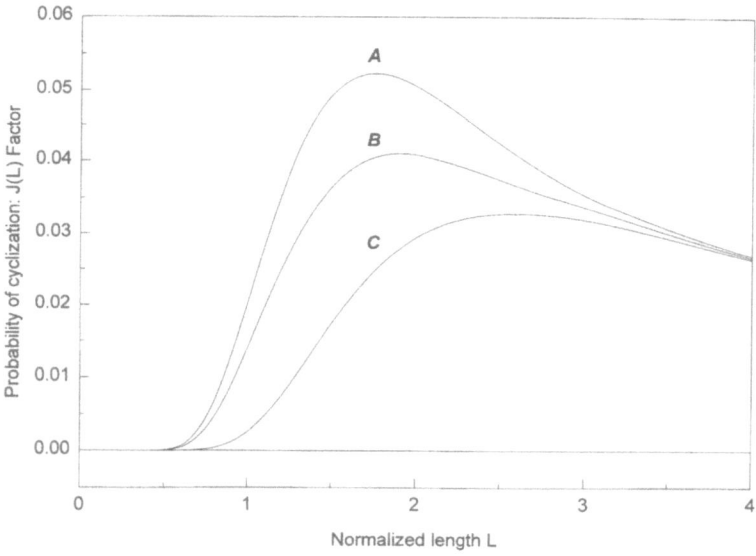

Fig. 4. Theoretical probability of circularization for a twisted wormlike chain. The curves represent the theoretical length-dependent, stiffness-dependent, and twist-dependent probability of ring closure for linear DNA molecules as calculated according to Eqs. (1), (37-39), (60), (69), (70), (73), (C1-C3) in Shimada and Yamakawa (1984). L indicates the reduced contour length of the chains, i.e., the ratio between the length of the Kuhn segment λ^{-1} and the contour length of the DNA (number of bp x 3.4 Å). The J(L) factor is the Jacobson and Stockmayer (1950) J factor, which is related to the ring closure probability by Eq. (1) in Shimada and Yamakawa (1984). The values of the Poisson's ratio σ and the fractional twist r are: A, $\sigma = -0.2$, $r = 0$; B, $\sigma = 0.8$, $r = 0$; C, $\sigma = -0.2$, $r = 0.5$. A and C correspond to the theoretical curves for random B-DNA (Shimada and Yamakawa 1984; Bacolla et al. 1997) and indicate the upper and lower boundaries of the twist-dependent J factor. With $\lambda^{-1} = 950$ Å (persistence length of 475 Å) for random B-DNA, $L = 1$ corresponds to 279 bp (950 Å / 3.4 Å) for curves A and C

3.3.2
Cyclization Kinetics on DNA Fragments Containing (CTG·CAG)$_n$ and (CGG·CCG)$_n$

Linear DNA fragments with uninterrupted (CTG·CAG)$_n$ or (CGG·CCG)$_n$ triplet repeats were constucted for CK experiments (Bacolla et al. 1997). Bimolecular and cyclization events were scored by sealing the hybridized ends with T4 DNA ligase in reaction mixtures where the radioactive fragments were reacted with the enzyme for varying time. These samples were electrophoresed through polyacrylamide gels to separate the linear monomers, the linear dimers, and the circles. The concentration of each species was calculated, and the molar J factor was computed for each fragment. The molar J factor corresponds to the ratio of the first-order kinetics of cyclization to the second-order kinetics of bimolecular reaction (Bacolla et al. 1997) and is the primary quantity needed to derive the elastic forces. The molar J factors were then plotted as a function of length, and the formulas used to construct the curves in Fig. 4 were applied so as to find the best fit for the experimental data. The results are reported in Table 3 along with the values determined previously for DNA of random sequence (Shore et al. 1981; Shore and Baldwin 1983a; Shimada and Yamakawa 1984; Bacolla et al. 1997). The most significant result is that the bending moduli for the TRS were much lower (approximately 40 %) than for random DNA. We interpreted this finding to mean that DNA composed of (CTG·CAG)$_n$ or (CGG·CCG)$_n$ bends more easily than the bulk of chromosomal DNA. On the other hand, both the torsional forces and the helical repeats (i.e., the number of base pairs per turn of double helix) of these TRS were identical to those of DNA of random sequence.

Table 3. Elastic constants and helical repeat for (CTG·CAG)n and (CGG·CCG)n

Parameter	Bending modulus α (x10^{-19} erg·cm)	Torsional modulus β (x10^{-19} erg·cm)	Helical repeat h^0 (bp/turn)
Random DNA	1.92	2.4	10.46
(CTG·CAG)$_n$	1.13	2.3	10.41
(CGG·CCG)$_n$	1.27	2.4	10.35

The bending modulus α, the twisting modulus β, and the helical repeat h^0 were derived from the values of Poisson ratio σ, the length of the Kuhn segment λ^{-1}, and constant torsion τ_0 used to fit the molar J factors. The values were: (CTG·CAG)$_n$, $\sigma = -0.51$, $\lambda^{-1} = 556$ Å (which corresponds to a persistence length of 278 Å), $\tau_0 = 0.1775$; (CGG·CCG)$_n$, $\sigma = -0.46$, $\lambda^{-1} = 630$ Å (which corresponds to a persistence length of 315 Å), $\tau_0 = 0.1786$. The conversion factors were: $\alpha = k_B T \lambda^{-1}$); $\beta = \alpha / (1 + \sigma)$; $h^0 = 2\pi / \tau_0 l_{bp}$; where k_B is the Boltzmann constant, T the absolute temperature, and l_{bp} the distance between adjacent base pairs, 3.4 Å.

In addition to bending, we pointed out the importance of supercoiling in general biological processes. The amount of stress that is required to supercoil a DNA molecule depends on its flexibility. Therefore, one would expect that the TRS supercoils more easily than bulk chromosomal DNA. Experimental support of this contention was also found. First, the electrophoretic migration of supercoiled plasmids was greatly influenced by the length of an inserted fragment when it contained a TRS sequence. Since the effect was not observed when fragments of random DNA sequence were cloned instead of the TRS, we concluded that the altered migration reflected changes in the topological shape of the plasmids caused by the TRS. Specifically, under the stress of supercoiling, the TRS were more contorted (with a greater writhe) than the rest of the plasmid. Second, circles of fragments containing $(CTG \cdot CAG)_{64}$, $(CGG \cdot CCG)_{70}$, and $(CGG \cdot CCG)_{71}$ (224–245 bp in length) migrated on polyacrylamide gels as two or more species, indicating that these molecules contained a different number of helical turns. Since topologically isomeric circles also form with random DNA, but only at longer lengths, the formation of multiple circular species at such short lengths must have been due to the greater flexibility of the TRS as compared to random DNA.

The bending and torsional moduli also enable the application of analytical equations that predict the energy required to supercoil DNA. These theoretical calculations yield a parameter (K) that expresses the amount of free energy stored at each base pair at every superhelical turn (Shimada and Yamakawa 1985). K was computed for DNA of $(CTG \cdot CAG)_n$ and $(CGG \cdot CCG)_n$ composition up to 10000 bp (Gellibolian et al. 1997). The results showed that the values for the TRS were lower than those for random B-DNA, confirming that less energy is required to supercoil a TRS than a random B-DNA. However, the differences in these free energies of supercoiling between TRS and random B-DNA varied with length. They first increased, reached a maximum at approximately 520 bp, and then gradually decreased to steady levels. This unexpected result reveals a close correspondence between the ability of the TRS to supercoil and the length of 180–200 triplet repeats (540–600 bp) that separates the premutation range from the full mutation range in the fragile X and myotonic dystrophy pathologies. Therefore, these analyses reveal that supertwisting the highly flexible and writhed TRS promotes biochemical events that contribute deleteriously to the stability of these sequences.

3.4
DNA Flexibility as a Source of Genetic Instability

Both the accumulation of supercoiling and the induced sequence instability may occur through multiple mechanisms. Figure 5 shows the preferential partitioning of writhe (supercoiling) within a TRS in chromosomal DNA. A linear DNA segment composed of 50 % TRS (left) and 50 % random B-DNA (right) is flanked by two protein complexes (rectangles). If the filled rectangle rotates while the open rectangle remains static, the TRS and B-DNA overwind or unwind, depending on the sense of rotation of B, and therefore writhe. However, since the TRS is more flexible than B-DNA, more writhe accumulates within the TRS than within B-DNA. Furthermore, the difference in partitioning of supercoiling depends on the length of the TRS and B-DNA, reaching a maximum when both TRS and B-DNA are 520 bp long.

Translocation of both RNA and DNA polymerase complexes induces overwinding and unwinding, and topoisomerases relieve or induce supercoiling (Wang 1996). Therefore, these activities may lead to the preferential accumulation of supercoiling within the TRS in the chromosome. Alternative base pair associations, such as slippage-associated hairpins, are favored for unwound DNA, and these may lead to genetic instability (Wells and Sinden 1993). The

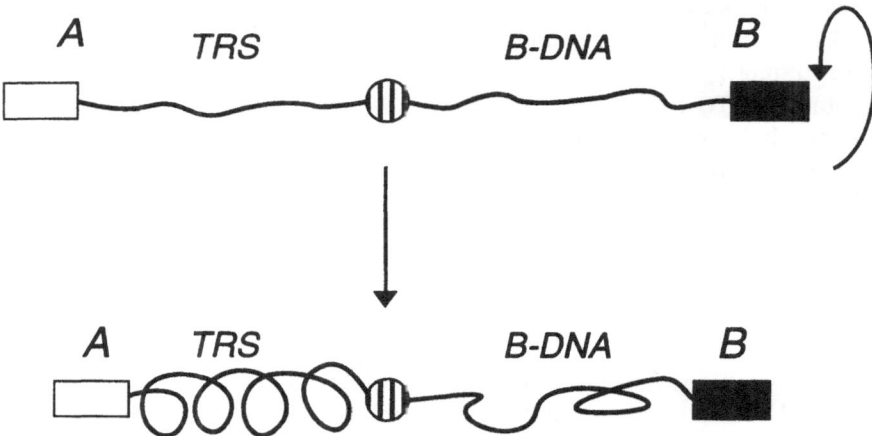

Fig. 5. The preferential partitioning of writhe within a region of DNA composed of random-sequence DNA adjoining a triplet repeat sequence. The two rectangles *A* and *B* represent protein complexes bound to the DNA, where *A* is static (its rotation is hindered), whereas *B* can rotate clockwise or counterclockwise (*top*). TRS [(CTG·CAG)$_n$ or (CGG·CCG)$_n$] and *B-DNA* indicate two segments of duplex DNA of equal length, and the *circle* represents a junction intersection that prevents supercoiling from diffusing between the two DNA segments. Rotation of *B* clockwise or counterclockwise overwinds or unwinds, respectively, the two DNA segments (*bottom*), whose helices adopt supercoils (writhe). However, since the TRS is more flexible than B-DNA, writhe will accumulate to a greater extent within the TRS region than within the B-DNA segment (Gellibolian et al. 1997). The quantitative calculations of the free energies of supercoiling were performed under the assumption that the DNA possesses only one bending modulus and one torsional modulus. This is obviously not the case when a TRS is embedded in random sequence DNA. The statistical mechanical calculations of such a mixed DNA are greatly complicated by the presence of two very different bending moduli, and we have not attempted such calculations at the present time. As a consequence, the two domains (B-DNA and TRS) are purposely separated. In addition, the conformation shown at the bottom is a schematic representation of the writhe which is not intended to differentiate between plectonemic and solenoidal or between right-handed and left-handed supercoiling

transcription-induced instability of the TRS is consistent with this mechanism (Bowater et al. 1997). In vitro, DNA polymerases stall during replication through long TRS, suggesting that these sequences form tertiary structures incompatible with polymerase progression (Kang et al. 1995b; Ohshima and Wells 1997). It is possible that the unremoved supercoils interfere with the progression of translocating protein complexes and favor the formation of alternative DNA interactions that lead to sequence instability. The large expansions observed in type 2 disease alleles and the size of approximately 180–200 repeats that demarkates the boundary between the premutation and the full-mutation range (Ashley and Warren 1995; Paulson and Fischbeck 1996) correlate with the increased ability of the TRS to supercoil at this length. Whether this correlation is significant with respect to the disease pathologies remains to be elucidated. Lastly, high levels of supercoiling bring multiple segments of DNA in contact with one another. Due to the repetitive nature of the TRS, these close proximities may induce interstrand interactions that lead to destabilization of duplex DNA and, ultimately, to genetic instability.

In summary, it is likely that the flexible and highly writhed duplex conformations of $(CTG \cdot CAG)_n$ and $(CGG \cdot CCG)_n$ play an important role in the strand slippage, hairpin-loop formation, DNA polymerase pausing, daughter strand hairpin formation, and possibly other events that are seminal to the expansion/deletion processes that give rise to anticipation.

4
Prospects for the Future

Enormous progress has been made in our understanding of the genetic basis of several human neurological diseases. In addition, we have been able to glean a great deal about the molecular mechanisms of the processes which are responsible for genetic instabilities from simpler systems. Future work will focus on simultaneous advances in both complex human genetic systems and bacteria. Furthermore, we may expect to see substantial progress with investigations on yeast and other lower eukaryotes as well as animal model systems such as transgenic mice. Unfortunately, the phenotypes of transgenic mice do not always correlate with those of human patients. Thus it is possible that the nature of mammalian development may be too complex to allow complete identification and molecular pathways, at least in the near future.

The fundamental nature of the processes involved (transcription, replication, repair) suggests that these instabilities are likely to be conserved through evolution, although the details may vary in different systems. The observation that more than one pathway produces instabilities in long TRS in *E. coli* (Bowater et al. 1997) may relate to different mechanisms for type 1 and type 2 disease. In all situations in which the expansion of a TRS is associated with a human disease, the TRS is located within a gene (Ashley and Warren 1995; Paulson and Fischbeck 1996). Thus there is potential for transcription-associated events to influence the stability of TRS in all disorders. The regulation of transcription is specific for each promoter, and this may provide one mechanism by which local phenomena alter the stability of a particular TRS in a manner that is dependent on its position.

Despite the optimism expressed above, numerous fundamental questions remain which will require innovative strategies for future experimentation. The degree of suffering encountered in patients with these diseases serves as an impetus for our best efforts to develop rational therapeutic strategies that will benefit mankind.

Acknowledgements. This work was supported by grants from the National Institutes of Health (GM52982) and the Robert A. Welch Foundation. The authors wish to thank our colleagues for providing preprints of manuscripts for this review, and we express our appreciation to our colleagues who performed the original experiments from this as well as our collaborator's laboratories. These individuals include the following: Drs. S. Amirhaeri, P. Chastain, E. Eichler, R. Gellibolian, D. Giedroc, J. Griffith, S. C. Harvey, A. Jaworski, S. Kang, S. Kramer, J. Larson, S. Levene, D. Nelson, K. Ohshima, B. A. Oostra, P. Parniewski, W. Rosche, M. Shimizu, R. Sinden, B. D. Stollar, and Y.-H. Wang.

References

Aaltonen LA, Peltomaki P, Leach FS, Sistonen P, Pylkkanen L, Mecklin JP, Jarvinen H, Powell SM, Jen J, Hamilton SR, Petersen GM, Kinzler KW, Vogelstein B, de la Chapelle A (1993) Clues to the pathogenesis of familial colorectal cancer. Science 260:812–816

Alfano C, McMacken R (1988) The role of template superhelicity in the initiation of bacteriophage λ DNA replication. Nucleic Acids Res 15:9611–9630

Armour JA, Jeffreys AJ (1992) Biology and applications of human minisatellite loci. Curr Opin Genet Dev 2:850–856

Ashley CT, Warren ST (1995) Trinucleotide repeat expansion and human disease. Annu Rev Genet 29:703–728

Ashley CT, Sutcliffe JS, Kunst CB, Leiner HA, Eichler EE, Nelson DL, Warren ST (1993) Human and murine FMR-1: alternative splicing and translational initiation downstream of the CGG-repeat. Nat Genet 4:244–251

Bacolla A, Gellibolian R, Shimizu M, Amirhaeri S, Kang S, Ohshima K, Larson JE, Harvey SC, Stollar BD, Wells RD (1997) Flexible DNA: genetically unstable CGG and CTG from human hereditary neuromuscular disease genes. J Biol Chem 272:16783–16792

Baker SM, Bronner CE, Zhang L, Plug AW, Robatzek M, Warren G, Elliott EA, Yu J, Ashley T, Arnheim N, Flavell RA, Liskay RM (1995) Male mice defective in the DNA mismatch repair gene PMS2 exhibit abnormal synapsis in meiosis. Cell 82:309–319

Bednar J, Furrer P, Katrich V, Stasiak AZ, Dubochet J, Stasiak A. (1995) Determination of DNA persistence length by cryo-electron microscopy. Separation of the static and dynamic contributions to the apparent persistence length of DNA. J Mol Biol 254:579–594

Bingham PM, Scott MO, Wang S, McPhaul MJ, Wilson EM, Garbern JY, Merry DE, Fischbeck KH (1995) Stability of an expanded trinucleotide repeat in the androgen receptor gene in transgenic mice. Nat Genet 9:191–196

Bowater RP, Rosche WA, Jaworski A, Sinden RR, Wells RD (1996) Relationship between *Escherichia coli* growth and deletions of CTG·CAG triplet repeats in plasmids. J Mol Biol 264:82–96

Bowater RP, Jaworski A, Larson JE, Parniewski P, Wells RD (1997) Transcription increases the deletion frequency of CTG·CAG triplet repeat sequences from human neuromuscular disease genes in *E. Coli*. Nucleic Acids Res 25:2861-2868

Burke JR, Enghild JJ, Martin ME, Jou YS, Myers RM, Roses AD, Vance JM, Strittmatter WJ (1996) Huntingtin and DRPLA proteins selectively interact with the enzyme GAPDH. Nat Med 2:347-350

Burright EN, Clark HB, Servadio A, Matilla T, Feddersen RM, Yunis WS, Duvick LA, Zoghbi HY, Orr HT (1995) SCA1 transgenic mice: a model for neurodegeneration caused by an expanded CAG trinucleotide repeat. Cell 82:937-948

Campuzano V, Montermini L, Molto MD, Pianese L, Cossee M, Cavalcanti F, Monros E, Rodius F, Duclos F, Monticelli A, Zara F, Canizares J, Koutnikova H, Bidichandani SI, Gellera C, Brice A, Trouillas P, de Michele G, Filla A, de Frutos R, Palau F, Patel PI, di Donato S, Mandel J-L, Cocozza S, Koenig M, Pandolfo M (1996) Friedreich's ataxia: autosomal recessive disease caused by an intronic GAA triplet repeat expansion. Science 271:1423-1427

Caskey CT, Pizzuti A, Fu YH, Fenwick RG Jr, Nelson DL (1992) Triplet repeat mutations in human disease. Science 256: 784-789

Charlesworth B, Sniegowski P, Stephan W (1994) The evolutionary dynamics of repetitive DNA in eucaryotes. Nature 371:215-220

Chastain PD, Eichler EE, Kang S, Nelson DL, Levene SD, Sinden RR (1995) Anomalously rapid electrophoretic mobility of DNA containing triplet repeats associated with human disease genes. Biochemistry 34:16125-16131

Chen X, Mariappan SVS, Catasti P, Ratliff R, Moyzis K, Laayoun A, Smith SS, Bradbury EM, Gupta G (1995) Hairpins are formed by the single DNA strands of the fragile X triplet repeats: structure and biological implications. Proc Natl Acad Sci USA 92:5199-5203

Chung M-y, Ranum LPW, Duvick LA, Servadio A, Zoghbi HY, Orr HT (1993) Evidence for a mechanism predisposing to intergenerational CAG repeat instability in spinocerebellar ataxia type I. Nat Genet 5:254-258

de la Chapelle A, Peltomaki P (1995) Genetics of hereditary colon cancer. Annu Rev Genet 29:329-348

DePamphilis ML (1993) Eukaryotic DNA replication: anatomy of an origin. Annu Rev Biochem 62:29-63

de Wind N, Dekker M, Berns A, Radman M, de Riele H (1995) Inactivation of the mouse Msh2 gene results in mismatch repair deficiency, methylation tolerance, hyperrecombination, and predisposition to cancer. Cell 82:321-330

Diekmann S, Hillen W, Morgeneyer B, Wells RD, Pörscke D (1982) Orientation relaxation of DNA restriction fragments and the internal mobility of the double helix. Biophys Chem 15:263-270

Dover G (1995) Slippery DNA runs on and on and on... Nat Genet 10:254

DuPraw EJ (1974) Quantitative constraints in the arrangement of human DNA. Cold Spring Harbor Symp Quant Biol 38:87-98

Dutch-Belgian Fragile X Consortium (1994) Fmr1 knockout mice: a model to study Fragile X mental retardation. Cell 78:23-33

Duyao M, Ambrose C, Myers R, Novelletto A, Persichetti F, Frontali M, Folstein S, Ross C, Franz M, Abbott M et al. (1993) Trinucleotide repeat length instability and age of onset in Huntington's disease. Nat Genet 4:387-392

Duyao MP, Auerbach AB, Ryan A, Persichetti F, Barnes GT, McNeil SM, Ge P, Vonsattel J-P, Gusella JF, Joyner AL, MacDonald ME (1995) Inactivation of the mouse Huntington's disease gene homologue Hdh. Science 269:407-410

Eichler EE, Holden JJ, Popovich BW, Reiss AL, Snow K, Thibodeau SN, Richards CS, Ward PA, Nelson DL (1994) Length of uninterrupted CGG repeats determines instability in the FMR1 gene. Nat Genet 8:88-94

Eichler EE, Hammond HA, Macpherson JN, Ward PA, Nelson DL (1995) Population survey of the human FMR1 CGG repeat substructure suggests biased polarity of the loss of AGG interruption. Hum Mol Genet 4:2199-2208

Eshelmann JR, Markowitz SD (1996) Mismatch repair defects in human carcinogenesis. Hum Mol Genet 5:1489-1494

Fishel R, Kolodner RD (1995) Identification of mismatch repair genes and their role in the development of cancer. Curr Opin Genet Dev 5:382–395

Fishel R, Lescoe MK, Rao MRS, Copeland NG, Jenkins NA, Garber J, Kane M, Kolodner R (1993) The human mutator gene homolog MSH2 and its association with hereditary nonpolyposis colon cancer. Cell 75:1027–1038

Freudenreich CH, Stavenhagen JB, Zakian VA (1997) Stability of a CTG/CAG trinucleotide repeat in yeast is dependent on its orientation in the genome. Mol Cell Biol 17:2090–2098

Freund AM, Bichara M, Fuchs RP (1989) Z-DNA-forming sequences are spontaneous deletion hot spots. Proc Natl Acad Sci USA 86:7465–7469

Fry M, Loeb LA (1994) The fragile X syndrome d(CGG)$_n$ nucleotide repeats form a stable tetrahelical structure. Proc Natl Acad Sci USA 91:4950–4954

Fu Y-H, Kuhl DPA, Pizzuti A, Pieretti M, Sutcliffe JS, Richards S, Verkerk AJMH, Holden JJA, Fenwick RG Jr, Warren ST, Oostra BA, Nelson DL, Caskey CT (1991) Variation of the CGG repeat at the fragile X site results in genetic instability: resolution of the Sherman paradox. Cell 67:1047–1058

Fu Y-H, Pizzuti A, Fenwick RG Jr, King J, Rajnarayan S, Dunne PW, Dubel J, Nasser GA, Ashizawa T, de Jong P, Wieringa B, Korneluk R, Perryman MB, Epstein HF, Caskey CT (1992) An unstable triplet repeat in a gene related to myotonic muscular dystrophy. Science 255:1256–1258

Gacy AM, Goellner G, Juranic N, Macura S, McMurray CT (1995) Trinucleotide repeats that expand in human disease form hairpin structures in vitro. Cell 81:533–540

Gastier JM, Pulido JC, Sunden S, Brody T, Buetow KH, Murray JC, Weber JL, Hudson TJ, Sheffield VC, Duyk GM (1995) Survey of trinucleotide repeats in the human genome: assessment of their utility as genetic markers. Hum Mol Genet 4:1829–1836

Gellibolian R, Bacolla A, Wells RD (1997) Triplet repeat instability and DNA topology: A expansion model based on statistical mechanics. J Biol Chem 27:16793–16797

Goldberg YP, Kalchman MA, Metzler M, Nasir J, Zeisler J, Graham R, Koide HB, O'Kusky J, Sharp AH, Ross CA, Jirik F, Hayden MR (1996) Absence of disease phenotype and intergenerational stability of the CAG repeat in transgenic mice expressing the human Huntington disease transcript. Hum Mol Genet 5:177–185

Hagerman PJ (1988) Flexibility of DNA. Annu Rev Biophys Biophys Chem 17:265–286

Hagerman PJ, Ramadevi VA (1990) Application of the method of phage T4 DNA ligase-catalyzed ring-closure to the study of DNA structure. I. Computational analysis. J Mol Biol 212:351–362

Hancock JM (1996) Simple sequences and the expanding genome. Bioessays 18:421–425

Harris S, Moncrieff C, Johnson K (1996) Myotonic dystrophy: will the real gene please step forward! Hum Mol Genet 5:1417–1423

Hodges-Garcia Y, Hagerman P (1995) Investigation of the influence of cytosine methylation on DNA flexibility. J Biol Chem 270:197–201

Huntington's Disease Collaborative Research Group (1993) A novel gene containing a trinucleotide repeat that is expanded and unstable on Huntington's disease chromosomes. Cell 72:971–983

Ikeda H, Yamaguchi M, Sugai S, Aze Y, Narumiya S, Kakizuka A (1996) Expanded polyglutamine in the Machado-Joseph disease protein induces cell death in vitro and in vivo. Nat Genet 13:196–202

Imbert G, Saudou F, Yvert G, Devys D, Trottier Y, Garnier J-M, Weber C, Mandel J-L, Cancel G, Abbas N, Durr A, Didierjean O, Stevanin G, Agid Y, Brice A (1996) Cloning of the gene for spinocerebellar ataxia 2 reveals a locus with high sensitivity to expanded CAG/glutamine repeats. Nat Genet 14:285–291

Ionov Y, Peinado M, Malkhosyan S, Shibata D, Perucho M (1993) Ubiquitous somatic mutations in simple repeated sequences reveal a new mechanism for colonic carcinogenesis. Nature 363:558–561

Jacobson H, Stockmayer WH (1950) Intramolecular reaction in polycondensations. I. The theory of linear systems. J Chem Phys 18:1600–1606

Jaworski A, Higgins NP, Wells RD, Zacharias W (1991) Topoisomerase mutants and physiological conditions control supercoiling and Z-DNA formation in vivo. J Biol Chem 266:2576–2581

Jaworski A, Rosche WA, Gellibolian R, Kang S, Shimizu M, Sinden RR, Wells RD (1995) Mismatch repair in *Escherichia coli* enhances instability in vivo of $(CTG)_n$ triplet repeats from human hereditary diseases. Proc Natl Acad Sci USA 92:11019–11023

Jeffreys AJ, Tamaki K, MacLeod A, Monckton DG, Neil DL, Armour JAL (1994) Complex gene conversion events in germline mutation at human minisatellites. Nat Genet 6:136–145

Kahn JD, Yun E, Crothers DM (1994) Detection of localized DNA flexibility. Nature 368:163–166

Kang S, Jaworski A, Ohshima K, Wells RD (1995a) Expansion and deletion of CTG triplet repeats from human disease genes are determined by the direction of replication. Nat Genet 10:213–218

Kang S, Ohshima K, Shimizu M, Amirhaeri S, Wells RD (1995b) Pausing of DNA synthesis in vitro at specific loci in CTG and CGG triplet repeats from human hereditary diseases. J Biol Chem 270:27014–27021

Kang S, Ohshima K, Jaworski A, Wells RD (1996) CTG triplet repeats from the myotonic dystrophy gene are expanded in *E. coli* distal to the replication origin as a single large event. J Mol Biol 258:543–547

Kawaguchi Y, Okamoto T, Taniwaki M, Aizawa M, Inoue M, Katayama S, Kawakami H, Nakamura S, Nishimura M, Akiguchi I, Kimura J, Narumiya S, Kakizuka A (1994) CAG expansions in a novel gene for Machado-Joseph disease at chromosome 14q32.1. Nat Genet 8:221–228

Kinzler KW, Vogelstein B (1996) Lessons from hereditary colorectal cancer. Cell 87:159–170

Koide R, Ikeuchi T, Onodera O, Tanaka H, Igarashi S, Endo K, Takahashi H, Kondo R, Ishikawa A, Hayashi T, Saito M, Tomoda A, Miike T, Naito H, Ikuta F, Tsuji S (1994) Unstable expansion of CAG repeat in hereditary dentatorubal-pallidoluysian atrophy (DRPLA). Nat Genet 6:9–13

Kolodner RD (1995) Mismatch repair: mechanisms and relationship to cancer susceptibility. Trends Biochem Sci 20:397–401

Krontiris TG (1995) Minisatellites and human disease. Science 269:1682–1683

Kunst CB, Warren ST (1994) Cryptic and polar variation of the fragile X repeat could result in predisposing normal alleles. Cell 77:853–861

LaSpada AR, Wilson EM, Lubahn DB, Harding AE, Fischbeck KH (1991) Androgen receptor gene mutations in X-linked spinal and bulbar muscular atrophy. Nature 352:77–79

Leach FS, Nicolaides NC, Papadopoulos N, Liu B, Jen J, Parsons R, Peltomaki P, Sistonen P, Aaltonen LA, Nystrom-Lahti M, Guan X-Y, Zhang J, Meltzer PS, Yu J-W, Kao F-T, Chen DJ, Cerosaletti KM, Fournier REK, Todd S, Lewis T, Leach RJ, Naylor SL, Weissenbach J, Mecklin JP, Jarvinen H, Petersen GM, Hamilton SR, Green J, Jass J, Watson P, Lynch HT, Trent JM, de la Chapelle A, Kinzler KW, Vogelstein B (1993) Mutations of a mutS homolog in hereditary non-polyposis colorectal cancer. Cell 75:1215–1225

Levinson G, Gutman GA (1987a) High frequencies of short frameshifts in poly-CA/TG tandem repeats borne by bacteriophage M13 in *Escherichia coli* K-12. Nucleic Acids Res 15:5323–5338

Levinson G, Gutman GA (1987b) Slipped-strand mispairing: a major mechanism for DNA sequence evolution. Mol Biol Evol 4:203–221

Li X-J, Li S-H, Sharp AH, Nucifora FC Jr, Schilling G, Lanahan A, Worley P, Snyder SH, Ross CA (1995) A huntingtin-associated protein enriched in brain with implications for pathology. Nature 378:389–402

Loeb LA (1994) Microsatellite instability: marker of a mutator phenotype in cancer. Cancer Res 54:5059–5063

Lustig AJ, Petes TD (1993) Genetic control of simple sequence stability in yeast. In: Davies KE, Warren ST (eds) Genome rearrangement and stability, vol. 7, Cold Spring Harbor Laboratory Press, Cold Spring Harbor, pp 79–106

Mahadevan M, Tsilfidis C, Sabourin L, Shutler G, Amemiya C, Jansen G, Neville C, Narang M, Barcelo J, O'Hoy K, Leblond S, Earle-Macdonald J, de Jong P, Wieringa B, Korneluk RG (1992) Myotonic dystrophy mutation: an unstable CTG repeat in the 3' untranslated region of the gene. Science 255:1253–1255

Mangiarini L, Sathasivam K, Seller M, Cozens B, Harper A, Hetheringthon C, Lawton M, Trottier Y, Lehrach H, Davies SW, Bates GP (1996) Exon 1 of the HD gene with an expanded CAG repeat is sufficient to cause a progressive neurological phenotype in transgenic mice. Cell 87:493–506

Marra G, Boland CR (1995) Hereditary nonpolyposis colorectal cancer: the syndrome, the genes, and historical perspectives. J Natl Can Inst 87:1114–1125

Mitas M, Yu A, Dill J, Haworth IS (1995) The trinucleotide repeat sequence d(CGG)$_{15}$ forms a heat-stable hairpin containing G^{syn}·G^{anti} base pairs. Biochemistry 34:12803–12811

Mitchell JE, Newbury SF, McClellan JA (1995) Compact structures of d(CNG)$_n$ oligonucleotides in solution and their possible relevance to fragile X and related human genetic diseases. Nucleic Acids Res 23:1876–1881

Modrich P (1991) Mechanisms and biological effects of mismatch repair. Annu Rev Genet 25:229–253

Modrich P, Lahue R (1996) Mismatch repair in replication fidelity, genetic recombination, and cancer biology. Annu Rev Biochem 65:101–133

Monckton D, Coolbaugh MI, Ashizawa KT, Sicilano MJ, Caskey CT (1997) Hypermutable myotonic dystrophy CTG repeat mouse transgenes. Nat Genet 15:193–196

Monckton DG, Wong LJ, Ashizawa T, Caskey CT (1995) Somatic mosaicism, germline expansions, germline reversions and intergenerational reductions in myotonic dystrophy males: small pool PCR analyses. Hum Mol Genet 4:1–8

Nagafuchi S, Yanagisawa H, Sato K, Shirayama T, Ohsaki E, Bundo M, Takeda T, Tadokoro K, Kondo I, Murayama N, Tanaka Y, Kikushima H, Umino K, Kurosawa H, Furukawa T, Nihei K, Inoue T, Sano A, Komure O, Takahashi M, Yoshizawa T, Kanazawa I, Yamada M (1994) Dentatorubral and pallidoluysian atrophy expansion of an unstable CAG trinucleotide on chromosome 12p. Nat Genet 6:14–18

Nasir J, Floresco SB, O'Kusky JR, Diewert VM, Richman JM, Zeisler J, Borowski A, Marth JD, Phillips AG, Hayden MR (1995) Targeted disruption of the Huntington's disease gene results in embryonic lethality and behavioral and morphological changes in heterozygotes. Cell 81:811–823

Nasir J, Goldberg YP, Hayden MR (1996) Huntington disease: new insights into the relationship between CAG expansion and disease. Hum Mol Genet 5:1431–1435

Ohshima K, Kang S, Wells RD (1996a) CTG triplet repeats from human hereditary disease are dominant genetic expansion products in E. coli. J Biol Chem 271:1853–1856

Ohshima K, Kang S, Larson JE, Wells RD (1996b) Cloning, characterization, and properties of seven triplet repeat DNA sequences. J Biol Chem 271:16773–16783

Ohshima K, Kang S, Larson JE, Wells RD (1996c) TTA·TAA triplet repeats in plasmids form a non-hydrogen bonded structure. J Biol Chem 271:16784–16791

Ohshima K, Wells RD (1997) Hairpin formation during DNA synthesis primer realignment in vitro in triplet repeat sequences from human hereditary disease genes. J Biol Chem 272:16798–16806

Orr HT, Chung MY, Banfi S, Kwiatkowski TJ Jr, Servadio A, Beaudet AL, McCall AE, Duvick LA, Ranum LP, Zoghbi HY (1993) Expansion of an unstable trinucleotide CAG repeat in spinocerebellar ataxia type 1. Nat Genet 4:221–226

Parsons R, Li G-M, Longley MJ, Fang W-h, Papadopoulos N, Jen J, de la Chapelle A, Kinzler KW, Vogelstein B, Modrich P (1993) Hypermutability and mismatch repair deficiency in RER$^+$ tumor cells. Cell 75:1227–1236

Paulson HL, Fischbeck KH (1996) Trinucleotide repeats in neurogenetic disorders. Annu Rev Neurosci 19:79–107

Pulst S-M, Nechiporuk A, Nechiporuk T, Gispert S, Chen X-N, Lopes-Cendes I, Pearlman S, Starkman S, Orozco-Diaz G, Lunkes A, DeJong P, Rouleau GA, Auberger G, Korenberg JR, Figueroa C, Sahba S (1996) Moderate expansion of a normally biallelic trinucleotide repeat in spinocerebllar ataxia type 2. Nat Genet 14:269–276

Reitmair AH, Schmits R, Ewel A, Bapat B, Redston M, Mitri A, Waterhouse P, Mittrucker HW, Wakeham A, Liu B, Thomason A, Griesser H, Gallinger S, Ballhausen WG, Fishel R, Mak TW (1995) Msh2 deficient mice are viable and susceptible to lymphoid tumours. Nat Genet 11:64–70

Richards RI, Sutherland GR (1992) Dynamic mutations: a new class of mutations causing human disease. Cell 70:709–712

Rosche WA, Jaworski A, Kang S, Kramer SF, Larson JE, Giedroc DP, Wells RD, Sinden RR (1996) Single strand DNA binding protein enhances the stability of CTG triplet repeats in *Escherichia coli*. J Bacteriol 178:5042–5044

Sanpei K, Takano H, Igarashi S, Sato T, Oyake M, Sasaki H, Wakisaka A, Tashiro K, Ishida Y, Ikeuchi T, Koide R, Saito M, Sato A, Tanaka T, Hanyu S, Takiyama Y, Nishizawa M, Shimizu N, Nomura Y, Segawa M, Iwabuchi K, Eguchi I, Tanaka H, Takahashi H, Tsuji S (1996) Identification of the spinocerebellar ataxia type 2 gene using a direct identification of repeat expansion and cloning technique, DIRECT. Nat Genet 14:277–284

Schleif R (1992) DNA looping. Annu Rev Biochem 61:199–223

Sherman SL, Jacobs PA, Morton NE, Froster-Iskenius U, Howard-Peebles PN, Nielsen KB, Partington MW, Sutherland GR, Turner G, Watson M (1985) Further segregation analysis of the fragile X syndrome with special reference to transmitting males. Hum Genet 69:289–299

Shimada J, Yamakawa H (1984) Ring-closure probability for twisted wormlike chains. Application to DNA. Macromolecules 17:689–698

Shimada J, Yamakawa H (1985) Statistical mechanics of DNA topoisomers. The helical worm-like chain. J Mol Biol 184:319–329

Shimizu M, Gellibolian R, Oostra BA, Wells RD (1996) Cloning, characterization, and properties of plasmids containing CGG triplet repeats from the fragile X gene. J Mol Biol 258:614–626

Shore D, Baldwin RL (1983a) Energetics of DNA twisting. I. Relation between twist and cyclization probability. J Mol Biol 170:957–981

Shore D, Baldwin RL (1983b) Energetics of DNA twisting. II. Topoisomer analysis. J Mol Biol 170:983–1007

Shore D, Langowski J, Baldwin RL (1981) DNA flexibility studied by covalent closure of short fragments into circles. Biochemistry 78:4833–4837

Sinden RR (1994) DNA structure and function. Academic Press, San Diego, California

Sinden RR, Wells RD (1992) DNA structure, mutations, and human genetic disease. Curr Opin Biotechnol 3:612–622

Smith GK, Jie J, Fox GE, Gao X (1995) DNA CTG triplet repeats involved in dynamic mutations of neurologically related gene sequences form stable duplexes. Nucleic Acids Res 23:4303–4311

Smith GP (1973) Unequal crossover and the evolution of multigene families. Cold Spring Harbor Symp Quant Biol 38:507–513

Strand M, Prolla TA, Liskay RM, Petes TD (1993) Destabilisation of tracts of simple repetitive DNA in yeast by mutations affecting DNA mismatch repair. Nature 365:274–276

Strand M, Earley MC, Crouse GF, Petes TD (1995) Mutations in the msh3 gene preferentially lead to deletions within tracts of simple repetitive DNA in *Saccharomyces cerevisiae*. Proc Natl Acad Sci USA 92:10418–10421

Sutherland GR, Richards RI (1995) Simple tandem DNA repeats and human genetic disease. Proc Natl Acad Sci USA 92:3636–3641

Tautz D, Schlotterer C (1994) Simple sequences. Curr Opin Genet Dev 4:832–837

Taylor WH, Hagerman PJ (1990) Application of the method of phage T4 DNA ligase-catalyzed ring-closure to the study of DNA structure. II. NaCl-dependence of DNA flexibility and helical repeat. J Mol Biol 212:363–376

Telenius H, Kremer B, Goldberg YP, Theilmann J, Andrew SE, Zeisler J, Adam S, Greenberg C, Ives EJ, Clarke LA, Hayden MR (1994) Somatic and gonadal mosaicism of the Huntington disease gene CAG repeat in brain and sperm. Nat Genet 6:409–414

Thibodeau SN, Bren G, Schaid D (1993) Microsatellite instability in cancer of the proximal colon. Science 260:816–819

Trinh TQ, Sinden RR (1991) Preferential DNA secondary structure mutagenesis in the lagging strand of replication in *E. coli*. Nature 352:544–547

Trottier Y, Lutz Y, Stevanin G, Imbert G, Devys D, Cancel G, Saudou F, Weber C, David G, Tora L, Agid Y, Brice A, Mandel J-L (1995) Polyglutamine expansion as a pathological epitope in Huntington's disease and four dominant cerebellar ataxias. Nature 378:403–406

Umar A, Kunkel TA (1996) DNA-replication fidelity, mismatch repair and genome instability in cancer cells. Eur J Biochem 238:297–307

Vogelstein B, Kinzler KW (1993) The multistep nature of cancer. Trends Genet 9:138–141

Wang JC (1996) DNA topoisomerases. Annu Rev Biochem 65:635–692

Wang JC, Davidson N (1966) On the probability of ring closure of lambda DNA. J Mol Biol 19:469–482

Wang Y-H, Amirhaeri S, Kang S, Wells RD, Griffith J (1994) DNA triplet repeats from the myotonic dystrophy gene are preferential nucleosome assembly sites in vitro. Science 265:669–671

Wang Y-H, Gellibolian R, Shimizu M, Wells RD, Griffith J (1996) Long repeating CCG triplet repeat blocks exclude nucleosomes: a possible mechanism for the nature of fragile sites in chromosomes. J Mol Biol 263:511–516

Warren ST (1996) The expanding world of trinucleotide repeats. Science 271:1374–1375

Warren ST, Ashley JCT (1995) Triplet repeat expansion mutations: the example of fragile X syndrome. Annu Rev Neurosci 18:77–99

Warren ST, Nelson DL (1994) Advances in molecular analysis of fragile X syndrome. J Am Med Assoc 271:536–542

Wells RD (1988) Unusual DNA structures. J Biol Chem 263:1095–1098

Wells RD (1996) Molecular basis of genetic instability of triplet repeats. J Biol Chem 271:2875–2878

Wells RD (1997) Triplet repeat diseases studied in man, microbes, and molecules. Am J Psychiatry 154:887

Wells RD, Sinden RR (1993) Defined ordered sequence DNA, DNA structure, and DNA-directed mutation. In: Davis K, Warren S (eds) Genome analysis, vol 7. Genome rearrangement and stability. Cold Spring Harbor Laboratory Press, Cold Spring Harbor, pp 107–138

Willems PJ (1994) Dynamic mutations hit double figures. Nat Genet 8:213–215

Zeitlin S, Liu JP, Chapman DL, Papaioannou VE, Efstratiadis A (1995) Increased apoptosis and early embyonic lethality in mice nullizygous for the Huntington's disease gene homologue. Nat Genet 11:155–163

Zoghbi HY (1996) The expanding world of ataxins. Nat Genet 14:237–238

Subject Index